D1265114

RESCUE
SQUAD

RESCUE SQUAD

by Larry Ferazani

WILLIAM MORROW & COMPANY, INC. NEW YORK 1974

Printed in the United States of America.

1 2 3 4 5   78 77 76 75 74

Library of Congress Catalog Card Number 74-14418

ISBN 0-688-00292-7

Book design: Helen Roberts

For my children;
and for the men of the Cambridge Fire Department

ACKNOWLEDGMENTS

I would like to give special thanks to my friend Dorian Fliegel, without whose invaluable help and guidance this book could not have been written.

I would also like to thank my editor, Nancy McGrath, for her continuing faith in this project, and Kathleen Fliegel for her generous assistance and unflagging encouragement.

=*ONE*

WE'D HAD A COUPLE OF FALSE ALARMS AND THEN a drunk who fell off a barstool around midnight up near Porter Square. The drunk had a slight gash in his forehead and we ran him down to Cambridge Hospital. Dave, Joe and I rode with him in the rear of the Rescue truck and he told us we were in for a lousy winter. His falling off the barstool was just the start of it. He said he was heading south for the winter. "Like the birds?" asked Joe. The drunk said he didn't give a shit about no birds. He said he knew a guy in real estate who was gonna get him into the land boom. The drunk was wearing a thin old sweater and he didn't have any socks on. It was a cold night and he was shivering and Davey threw a blanket over his shoulders.

Back at the house the only one on the floor was the man on watch; the other two companies that operate out of our house were both asleep. Billy Stone put on a fresh pot of coffee and Cooper came upstairs, and the five of us sat up for a while shooting the breeze. Around one o'clock we went down to sleep: Lieutenant Cooper to the officers' quarters, the rest of us to the Rescue Company dormitory. From the window beside my bed I could see the bare trees and the steeple of the church down the street in Harvard Yard.

Then the bells were coming in. Bodies bolted out of beds, hands fumbling for clothes in the blinding light. Men were running. The wall clock read two A.M. "Banks and Flagg streets," shouted Joe Finnegan over the clanging of the bells. He grabbed the pole and dropped. Billy Stone hit it next. Then Dave. Then me. The other companies were coming down their poles and scrambling for their apparatus. Everything was going.

The Rescue truck cut down Quincy Street. We pulled on boots, clipped coats shut, put on air tanks, helmets and gloves. Billy Stone shot the one-way on Banks, and we saw the smoke hanging in the street. Engine One was right behind us. Lieutenant Cooper was on the radio. Dave Swanson gave his helmet one last tug and reached for a halligan bar.

The Rescue stopped just short of the three-story frame house. Flames were pouring out of the second-floor front windows, shooting up into the night, spraying sparks. People stood on the opposite sidewalk, their faces orange in the light of the fire. As we leapt from the rear of the apparatus somebody screamed over the roar of the fire, "There's a baby trapped!" I wheeled around and saw a woman burying her face in her hands. Then we were running.

We hit the back door, pausing inside the hallway to put on our masks. We opened up the regulators and the safety bells rang, signaling that the air had started to flow from the tanks on our backs. The second-floor landing was hazy; we could see the smoke seeping out from under the door. Dave Swanson pulled off a glove and felt the door up and down for heat; it wasn't hot. Joe Finnegan pushed past us, climbing for the top floor. Dave jammed the halligan in at the lock, pulled back and popped the door. Crouched low against the wall, he turned the knob and let the door swing open. A wall of black smoke tumbled down on us.

We crawled forward on our hands and knees, groping frantically, totally blind in the thick smoke. Up ahead the fire crackled. A pain flashed through my shoulder and something rolled, crashed and shattered. My hand banged against metal; I felt a handle and knobs—we were in the kitchen. Dave opened a window. I moved forward and could feel the heat now. I got down low, close to the floor, where the smoke is lightest. Straight ahead I could see the red.

To the left we pressed through one doorway, then another. Dave was right behind me. My head snapped back as my helmet slammed into a solid wall. I reached out and felt a ledge that dropped off into the deep hollow of a tub. "Bathroom," I yelled through the mask.

There was a distant banging and shouting: Engine One was coming up the front stairway with the line. We found a room to the rear. Dave stopped and I shoved past him and lunged for the bed. It was a question of seconds: no one could survive in this deadly smoke. My arm swept over the mattress, then underneath the bed. Nothing. My hands ran along the walls, over shelves, under a desk, anywhere a crib might fit, a child hide, a body fall. A chair scraped and I reached across it and felt a window. It was stuck. I took it out cleanly with my helmet. As the glass crashed below I heard voices. Dave bumped into me. "Nothing," he shouted.

Somebody else grabbed me by the boot. "Who's this?" he yelled. I recognized Finnegan's voice.

"Larry."

"OK," he said. "Top floor's clear—smoke but no fire. Six is coming up the back. I'm going to help." Then he was gone.

Dave moved slowly up the hallway toward the fire, looking for another room. Flames were shooting out ahead, dancing in front of our eyes, crackling and snapping, lapping up over the doors. I could feel the heat drying the sweat off my face as I crawled behind him. My knees were aching and my head started to throb. I had strapped the mask too tight.

The wall of heat pressed against us and the weight of the tank was becoming uncomfortable. I felt suddenly as if I were going to black out. I sucked in deep on the air and sank down so low my mask touched the floor. Dave Swanson crept forward. He knew from the shape of the house that there had to be another room. I didn't see how he could stand the heat. In my mind the woman down in the street screamed again. Where the hell is Engine One, I thought. Where the hell are the Ladder Companies? I couldn't hear a thing above the roar of the fire. It was practically on top of me, searing my ears and the back of my neck. It was impossible to advance against that wall of heat. If the fire got behind us, if the hallway went up, we'd be cut off. If the ceiling caved in or the floor fell, if anything at all happened now, we were trapped without a line.

I tugged at Dave's boot as he crawled forward. He sat up and then fell forward. He's had it, I thought, and I reached out and he was gone. He shouted. I hurled myself after him into a room to the left. Glass shattered as he knocked out the window, and the heat slackened instantly. I found the bed and felt over and under it, then along the walls.

"Nothing," I shouted.

"Let's beat it," he shouted back.

We were moving for the door when I heard a faint cry. The baby. My heart began to pound: the last thing you want to find is what you're looking for. I hollered and threw myself at the bed, and we began the search again. We were like madmen, clawing and pulling at everything we touched: shoes, bedclothes, wires, lamps. There was a chest of drawers that stuck out from the wall. I reached in behind it. My hand recoiled as it touched flesh. I pulled my glove off. The fear drained out of me as I felt fur, and a pair of tiny claws dug into me.

"It's a cat," I shouted. "A goddamn fuckin' cat."

"Christ," said Dave. "Let's get the hell out of here."

We could hear the crashing of water tearing into the front rooms now. Engine One must be hitting the fire, I thought. I picked up the cat and we crawled back to the kitchen. Six was in there with an inch-and-a-half line. They didn't have masks on and they were all coughing. Somebody was choking bad. Bob Yanski had the tip. I got up close enough to kiss him; he was on his knees vomiting. Finnegan held him from behind. "Take the line," Joe shouted. "I'll get him out of here."

"Take this, too," I said, and handed him the cat.

Dave grabbed the tip and the stream of water sputtered and roared as he opened up. He advanced toward the front rooms. I crouched right behind him, feeding him the line, and two guys from Six backed us up, jockeying the line and coughing all the while.

Dave played the hose on the red and it darkened down. Now there wasn't much red left. "That's good," I was telling him. "That's good, Davey." We kept moving forward, looking for more fire but finding less and less.

The smoke cleared fast and we found ourselves in the charred remains of the living and dining rooms, face to face with Lieutenant Cooper and Billy Stone. They had ended up with Engine One's line. They looked awesome in their long coats and boots, their faces hidden by their helmets and masks. We pulled off the masks and the first whiff of the stinking air nearly turned my stomach. In one corner a TV set had melted down into a grotesque lump. Steam was rising from the debris-strewn floor and water dripped from the ceiling and stood in puddles. An army of guys sloshed through, playing their lights over the walls.

Deputy Chief Simmons, a big man dressed all in white, surveyed the damage. He was in charge of the entire operation, and responsible only to the Chief of the Department. "Take it out here, will ya, Martin?" he said. Jerry Martin of

Ladder One stepped forward and began to hit the wall with his ax. The Deputy set another truckman, John Riordan, to work on a corner of the ceiling. The Deputy observed carefully, watching for any hidden signs of fire. As Riordan ripped down the first chunk of plaster with his hook, a wisp of smoke escaped. The Deputy told him to take out the whole corner; after twenty-five years on the Department he knows pretty well at a glance where a fire might have gone.

Frank Mondello, Ken Hovey and Eddie Kilroy wrestled with the remains of the sofa; all that was left of it were the metal springs. "Hey, Dep, look at that," said Kilroy. The Deputy nodded. The couch looked like the source of the fire; probably nothing more than a burning cigarette had got it going. Two rooms were lost, but the building had been saved, and nobody was hurt.

Cooper signaled to us and we went down to take a break. The November air tasted clean and sharp as we emerged from the house. The street was filled with the flashing lights of the apparatus and the crackling static of their radios. There were lines everywhere and water flowing in the gutters. Spectators milled around on the sidewalk looking cold and disappointed.

We cleaned the facepieces of our masks and changed the air tanks and stored them away. Our faces were streaked with soot and our hair was matted down from the helmets. Too tired to drain the water from our boots, we sat knee to knee in our sweat-soaked clothes in the back of the Rescue and lit cigarettes. We watched while Joe carefully bandaged a finger cut by flying glass. Joe is a short, stocky, serious guy with a mustache and narrow eyes that screw up tight at the corners whenever he laughs. He's an expert on many subjects and he concentrates hard on whatever he does, and right then he was bandaging his finger.

"Want me to put ya in for a citation, Finnegan?" said Cooper.

"You can't write somebody up for cuttin' themselves on a can of beer," Davey said.

Cooper grinned. "We could say the injury was sustained while Private Finnegan attempted at considerable risk to his bodily person to douse the fire with all available means."

"The Chief'd say he should've used water," said Billy Stone.

"Tell him it was a chemical fire," I said.

"It wasn't no can of beer," Joe said, turning his finger slowly and inspecting his work. "It was a tin of imported smoked oysters—the kind that goes for three bucks down at Cardullo's. There was a nice dry white wine to go with 'em. You gotta have a nice dry white wine with oysters, ya know," he said.

"I didn't know that," Dave said.

"Oh, yeah," Joe said. "That's the only way."

"Well," said Dave, "what were ya doin' up there so long? Lettin' it age?"

"There was plenty of smoke up there," Joe said.

"There was plenty of smoke where we were," Dave said. "And fire, too." He turned to Cooper. "Where the hell were you guys with the line? It was hot as a bastard."

"Whaddaya mean?" said Cooper. "We were there. You didn't jockey any line tonight. You were chasing pussy cats."

Everybody laughed but Dave.

"Who was that woman?" I asked. "The one who—"

"Who the Christ knows?" Dave said with disgust. His face reddened and he lowered his head and drew the back of his arm across his perspiring forehead. Dave is a big, easygoing guy with blond hair and a boyish toothy grin. He's always ready to make a joke. But he had taken a real chance, and the tension hadn't quite left him yet. He had five years on the Rescue and it wasn't the first time somebody in a crowd had yelled that someone was trapped. It happens often enough. People get excited and rumors spread, or maybe, just every once in a while, somebody thinks he's being funny. But every single time you hear that cry, you push yourself that little bit extra. And Dave had stuck his neck way out. For nothing.

When I first came on the Department I thought the tough thing was pulling people out. That was what the people and the newspapers and the city officials and the public made a big deal about. But pulling people out is easy enough. If they're alive it isn't that much of a fire. If they're dead there isn't that much you have to do. The tough thing is going into a burning building. There's danger every time, and you have to do the same job whether or not there's anybody inside. It's when you can't find them, and have to keep on looking because somebody has made you believe they're in there or even just might be there, that it gets roughest.

"Well," said Billy Cooper when we had finished the cigarettes, "we'd better show our faces." We put our goods back on and trudged inside to give them a hand with the overhaul. The crew from Ladder One was hard at work pulling walls and ceilings, shoveling out debris and tar-papering the broken windows. They looked like coal miners with their shovels and lights and the whites of their eyes standing out in their soot-blackened faces. "OK on the Rescue," said the Deputy with a wave of his hand when he saw us come in.

"Beautiful," said Billy Cooper as we trooped back down the stairs.

We rode slowly back to quarters, stopping for the red lights. Harvard Square was all but deserted. A pair of taxi-cabs stood waiting, their drivers reading newspapers. Yellow lights showed in a couple of windows in the dormitories in the Yard, and I thought how good it was going to feel to wash and lie down.

We turned onto the wide apron in front of the house. Joe jumped down from the rear step and guided Billy Stone as he backed into our bay. None of the other companies were in yet, and the cavernous apparatus floor with its dull, brown-tiled walls looked deserted and cheerless.

The air brakes hissed and died as the truck jerked to a halt. Dave climbed down and plugged in the battery charger. Cooper went upstairs to write his report for the permanent records. From the patrol room at the rear of the floor Joe phoned Fire Alarm Headquarters, in the annex next door, to report our return. Dave and I took turns drinking from the water cooler, and Billy Stone went out to his car to look for a spare pack of cigarettes. He came back empty-handed. "Jesus," said Joe, heading for the Coke machine, "I gotta get me a tonic. Who feels like a tonic?" No one else did.

Dave, Billy Stone and I started up the stairs past the darkened room where the coats and helmets of the other shifts hung in rows like a battalion of scarecrows. The smell of smoke was heavy on us, and the stairs seemed endless.

A single ring of the phone cut through the stillness of the house. Joe's voice came over the loudspeaker: "*Attention. Rescue going out.*"

We hit the poles and ran to the apparatus. The call was for an automobile accident on Huron Ave.

The siren blared as we roared out around the Common and up Concord Ave., the truck pounding hard over the bumps and all the equipment rattling. Joe checked his watch.

"What time?" Dave asked.

"Four-twelve."

"It's probably some drunk," I said.

"Yeah," said Joe. "At this hour it's gotta be some drunk."

As we turned down Huron Ave. we saw the flickering blue lights of the police. We pulled up just beyond the wreck. The car had smashed head on into a tree. The front end was completely crushed. Glass littered the ground and radiator coolant was spreading out from underneath the hood. One of the cops shouted that the doors were stuck. The air smelled of gasoline. Cooper yelled to Billy Stone to radio for an Engine Company.

We looked through the shattered windshield. Inside, there was a boy and a girl. The guy was slouched against the steering post, his head and chest a mass of blood. The girl lay slumped against the door. Cooper and Swanson worked on the girl's side, Joe and I on the boy's. The door had buckled. Joe wedged in the bar and pulled on it, but the door was still stuck fast. Through the cracked side window I could see the blood dripping steadily from the guy's ear. Joe took another bite with the bar. We heaved back on it, grunting with the strain. Billy Stone added his weight and the door ripped free.

I leaned in over the guy and pushed back the seat. His eyes opened and he looked at me. He seemed young, not more than sixteen or seventeen. It was hard to tell with all the blood. The top of his head was split wide open. I ran my hands up each leg. The right ankle bulged and I could feel the unevenness of the flesh. I unzipped his jacket and felt his arms, chest and back. He winced as I touched his right side.

Next to me Cooper was examining the girl. She was just a kid. There was a deep gash over one eye and her arms were twisted back under her, like a rag doll. Suddenly Cooper turned to Swanson. "Get the ambu bag," he said. "Get the board." Then he swooped down over her and started compressing her chest and breathing into her mouth as best he could where she lay. Her heart had stopped.

Joe went for the spine board and a splint. I fastened a cervical collar around the boy's neck and began to dress his head wound. His eyes brimmed with tears and he moaned as I wrapped the cloth around him.

"You'll be all right," I told him. "We'll take care of you."

He shut his eyes tight.

"Where does it hurt? Can you tell me where it hurts?"

He shook his head slightly and bit his lip. "It don't hurt me," he said.

Cooper and Billy Stone lifted the girl out of the car. They laid her on the stretcher and Swanson slipped a plastic airway into her throat and placed the facepiece of the ambu bag over her mouth. Billy Stone took up the cardiac massage. Swanson squeezed the bag, forcing air into her lungs. Her face was chalk-white. The boy tried to look.

"Don't move," I told him. "We're taking care of her. She's all right."

"No," he said.

"Don't worry about her. You're going to be all right."

"I'm all right," he repeated, looking at me as if I wasn't there.

Joe slid in from the passenger's side and we placed the splint along the length of the boy's leg. He cried out when he felt the straps tighten. Joe slipped the board behind his back and we strapped him to it and fastened the collar around his neck. That way if his spine was broken there was less risk of severing the cord when we moved him.

"I'm gonna die," he said.

"Course not," said Joe.

"I'm gonna die."

"Nobody's gonna die," Joe told him. "That's crazy talk."

"I'm gonna die," he said. "With her."

He started to cry. Tears streaked down his face. Joe glanced at me. Something was all wrong. Maybe the kid had seen the girl before we got there. Maybe he knew. But it didn't make sense. He had to have been drinking to have wrecked the car that badly. But he wasn't loose like a drunk. And no drunk sits in a car and cries over an accident.

"We're going to move you now," I told him. "Don't you try to move. Let us move you."

We turned him and lifted him onto the stretcher. As we carried him to the Rescue, Engine Eight, who were standing by with their line laid, moved in to wash down the street.

Inside the Rescue, Cooper and Swanson were working over the girl on the stretcher bench. We squeezed past them and laid the boy on the deck, alongside the motor mount. The siren came on and Billy Stone pulled out. I placed the oxygen mask over the boy's face while Joe radioed Fire Alarm to alert Cambridge Hospital. He described the situation, speaking softly into the microphone so that the kid couldn't hear.

Dave Swanson was bent over the girl, compressing her chest while Cooper ventilated with the ambu bag. They would keep it up all the way in. They weren't doctors, and they couldn't stop now until they got to one, but they knew it was futile. They looked odd working so furiously over the girl with their faces still blackened from the fire. She was a very young girl with short brown hair and freckles, and except for where the blood had caked above her eye there wasn't a mark on her, and you could think she was just asleep.

The kid sobbed to himself. Jagged bits and pieces of glass clung to his blood-soaked jacket, and I noticed that there was blood on my jacket, too, and that my hands where they lay on his shoulders were thick with it. The kid stirred and strained against the straps. I told him to lie still, that we were almost there.

"I wanna die, too," he said, and as he spoke a little blood trickled from the corner of his mouth.

"She isn't dead," I told him.

He closed his eyes. "She is," he said. "She has to be."

I wiped the blood from his mouth. "Try not to talk," I said.

"I wanna die," he cried.

The siren was screaming as we shot through the tunnel onto Cambridge Street past the firehouse. It was really only a few more seconds now.

The kid's eyes opened and looked around and came to rest on me. "Why couldn't you let me die?" he said.

Joe heard it, too. We stared at each other. Suddenly it all made sense: the late hour, the high-speed, dead-center crash into the tree, his tears. Joe knelt beside the kid.

"Why did you do it?" he asked.

"Vicky got pregnant," the kid said. "I got her pregnant and we couldn't tell our folks." His chest heaved. "Why couldn't you let me die?" he wept.

We said all the junk we have to say, about how it's going to be OK, don't worry. The words sounded like lines from a script and while we were saying them we were wondering how these kids could have felt so trapped, how this thing could have happened—knowing only that, like all the other incredible things people do to themselves and each other in this city, it had happened. I looked at the girl, lying so peacefully as they worked over her, and that old feeling of helplessness and disbelief pushed aside the anger and pity. No matter how many times you've seen it, no matter how hard you try, you can't comprehend it. This girl had been alive; now she was dead.

We pulled up in front of the emergency entrance. Cooper and Swanson kept working on her as we carried her in. People ran ahead of us down the corridor into the trauma room. They continued cardiac massage and set up an EKG on her, and finally even gave her a shot of adrenaline to the heart, but there were no vital signs.

We went back out and brought the kid into the Emergency Room and laid him on one of the beds. As the intern leaned over him with a stethoscope one of the nurses drew the curtain.

On the way back to the house Joe said, "If you wanted to die, if you felt you hadda die, there's lots of easier ways to do it. You could just go in the garage, plug up your exhaust pipe and close the door. I had a couple of those cases."

Dave shrugged his shoulders.

"There's lots of ways," Joe said to me. "You know?"

"I guess so," I said, but I didn't feel much like talking about it then. First you take a chance in a fire for somebody who isn't there. Then the one guy you do help wants to die. That's the kind of night it was.

═ *TWO* ═══════════════

THIS CITY IS A BATTLEGROUND. SOME PEOPLE know it by its bookstores and its barrooms; I know it by corners where blood stains the pavement and streets where the smell of smoke hangs in the air.

My name is Larry Ferazani. I was born in Cambridge and I've lived here all my life. But I never knew this city for what it was until I joined the Rescue.

The Cambridge Fire Department Rescue is one of the busiest rescue companies in the country: last year we made nearly forty-seven hundred runs. It is also on of the few full-time companies that do both fire rescue and emergency medical service. Four squads of five men each, on a rotation schedule of ten-hour days and fourteen-hour nights, man the Rescue around the clock. Along with Engine Company One and the Aerial Tower (a specialized ladder truck capable of raising a bucket containing men), we work out of a firehouse located in the middle of Harvard University. Our house serves as administrative headquarters for the nine other houses in the Cambridge Fire Department. By all accounts we're good at what we do; articles and citations attest that we're one of the outstanding rescue companies in the United States.

The first half of our mission consists of fire fighting. We see more action than any of the other twelve Cambridge companies because, while each of them covers a specific district, we respond to every alarm in the entire city. Our primary responsibility at a fire is lifesaving: asphyxiation by smoke causes the majority of fire deaths, and we must get in and search every foot of a fire site where a victim might be lying overcome. We ventilate—open or break windows—as we go. When we have found and evacuated the people, or made absolutely sure there's no one inside, then we help the engine companies on the lines. If there are any victims, or if any of our men are injured, we administer first aid and transport them to the hospital.

The other half of our mission consists of providing emergency service to a hundred thousand people packed into one of the most densely populated cities in the country. The tasks we face are as varied as the city. When you mention Cambridge, people think of Harvard and MIT, two of the world's greatest universities; but this is also a town where 25 percent of the adult population have less than a high school education. When you mention Cambridge, people think of Harvard Square, a mecca for the young with its shops, cafes and throngs of students; but there are nearly as many people over sixty-five living within the vicinity of the Square as there are college students. Often when you mention Cambridge people think of historic Brattle Street, lined with the mansions of the old, wealthy Yankee families. They don't associate this town with the urban problems that plague other cities. But in sections of East Cambridge as many as half the housing units lack some or all of the basic plumbing facilities. Fifteen percent of the households exist below the poverty line, and serious crime is up 28 percent from last year, making Cambridge eighth in the nation in serious crime.

Cambridge is now 7 percent black, 7 percent Portuguese and over 5 percent Spanish-speaking. In East Cambridge we have all the problems and tensions that economic deterioration, rundown housing and an influx of ethnic minorities can create. Cambridge is a city of contrasts, and the emergencies we handle run from the street fights of the poor at one end of town to the suicides of the most privileged at the other.

The job we do takes its toll. Fire fighting is the most hazardous occupation in this country: nearly four out of every ten men are injured in the line of duty each year. And last year alone approximately one hundred and seventy-five firemen died fighting fires, a rate even higher than that of policemen killed.

But these are only statistics. Behind them, each of us sees the faces of a few friends. We don't talk about it much. Nobody forced us to take the job, and nobody's preventing us from quitting. But that doesn't keep us from seeing a hundred daily reminders that one day, any day, we might not be coming home. It doesn't keep us from worrying about what will happen to our families if we're not there.

On the Rescue there is another, more telling burden. You can't do this work day in and day out without having it get to you in one way or another. The accidents, the medical emergencies, the freak mishaps—these you can accept. It's the sick things that get you: the things people do to each other with guns, knives and hatred, and the things that drink, drugs and desperation do to people. Rescue work is lifting the roofs off people's houses and stepping for a moment into their lives; it's getting a good close look at the traps that have caught them and that they'll probably go right back to if they survive. It's seeing the raw human waste that the social machine produces through poverty, pressure and personal isolation. It's week after week of picking up casualties in a fight that can't be stopped. Admittedly, ours is an abnormal, twisted, distorted, crazy view of life; but it's the one you get, and after a while no other perspective seems quite as real. By any standard of reckoning, Cambridge is a favored city. But it's terrifying to contemplate what that says about our society.

Last month we had a call to oo down to one of the housing projects.This project is one of those dreary, red-brick, high-rise places where they stack people on top of one another. The slab of asphalt that serves as the courtyard is littered with glass from broken windows. The hallways are cluttered with rubbish and permeated with an odor of urine and human feces. We dread going down there. The residents are easily excited and routine jobs can become dangerously complicated.

That night last month we got to the project just after midnight. We found the kid we'd been sent for as soon as we opened the front door: he was lying on his belly headfirst down the stairs. There was a coagulated pool of blood around his head. The blood was running from his mouth, and his throat made a rasping sound. We cut away the tattered remains of his shirt. He had been butchered. His back was laced with knife wounds and each of the wounds had been ripped. He'd been carved up like a side of beef. People were screaming and hollering to get him out, to stop the bleeding, to do something, but you can't just pick up somebody in that condition and run. We had to spend precious seconds forcing the crowd back out of the way.

He was a twenty-two-year-old kid and he died on the operating table. I don't know why he was stabbed. But no explanation could block out that image in my mind, or the realization that there is that much hate and viciousness loose in the city.

For days after a thing like that you can't walk a street without noticing the sound of footsteps behind you; you can't start up a flight of stairs without wondering what might be waiting up ahead. Your wife doesn't go out to shop, or your children out to play, that you don't swear to yourself you'll never again let them out of your sight. And the thought of all the decent, law-abiding citizens gives you no comfort, because you've also seen how much they can be counted on in a pinch.

Only last week an elderly Chinese woman was robbed and beaten in broad daylight on Mass. Ave. near Central Square. The guy walked up to her, grabbed her purse, knocked her down and kicked her as she lay screaming on the sidewalk. There were plenty of witnesses. After the attacker fled she lay there in front of a grocery store for twenty minutes before somebody took the trouble to call for help.

You begin to wonder what we're coming to—not in the slums of New York City, or on the South Side of Chicago, or even just across the Charles River in the tougher sections of Boston, but right here in a place like Cambridge.

You begin to wonder how much more you want to see, how much more you can take. You wonder what it's doing to you. On the one hand you can become mechanical in your work and indifferent to other people's pain; you survive but you become a poorer rescue man, because so many facets of the job require a sensitivity to others. On the other hand you can shoulder the burden until it comes between you and your family and turns the town into a nightmare where every other street corner reminds you of someone's suffering. You start to figure that you'd be better off getting out.

The Chief thinks no one should serve more than five years on Rescue. He has practical reasons. The Rescue is a young man's job. A fireman gets more fire-fighting experience on Rescue than on any other company. He's given paramedical training and gains experience in handling emergencies of every conceivable kind. Rescue is a crucible which upgrades the quality of the personnel throughout the Department, and the Chief wants as many men as possible to have had the experience. The Chief cites these practical reasons for rotating Rescue assignments. But most men who have served on the Rescue say simply that five years is enough.

Service on the Rescue Company is entirely voluntary. A rescue man earns no more than a regular fireman and he can ask for a transfer at any time. Yet of those who have survived the initial shock of the gore and the senseless tragedies, none that I know of has ever quit. After two years on the Rescue I find that the longer I ride, the harder it is to stop. If the costs are increasingly high, it's also true that no other work I might do could provide as great a personal reward. I'm thirty-one years old now and for the first time in my life I have a calling which gives me deep satisfaction.

But I don't know. The work has a way of building up in you, and now another New England winter is closing in on us. This is the worst time of year for the Rescue. There are more fires, and they are more difficult and dangerous to fight. Winter doesn't officially begin for another week, but for us it began last night.

Our tour of duty started at five-thirty. We lined up for roll call behind the Rescue truck, our squad in front of the one that was coming off duty. Then the day shift was dismissed, Cooper read the general orders from the Chief of the Department on administrative and procedural matters, and we went upstairs to the locker room to change from our dress uniforms into our work clothes. Right away we had two quick medical runs: an asthmatic, then a diabetic.

At seven o'clock we had a little girl trapped in an elevator in an apartment building on Harvard Street. The elevator was stuck just slightly above the ground floor, and when we pressed our ears to the door we could hear her crying inside. She didn't respond to our questions. The manager of the building, a short man with a potbelly and the stub of a cigar jutting from the corner of his mouth, said he had sent for the repairman. He didn't want us breaking down his door.

"She's a colored girl," he told us. "She's coming in here all the time to ride the elevator."

Dave and Billy Stone shook the door, trying to spring it open, while Joe inserted the bar into the upper corner.

"You're not gonna break it in?" said the guy.

"We'll try to release it," Cooper said. "If that doesn't work we'll have to force it open."

"I don't want ya breakin' down my door."

"We can't wait around all night," Cooper said. "The girl's scared and we don't know what kind of shape she's in."

"She's just a colored girl that comes in here to ride my elevator," said the manager, carefully taking the cigar out of his mouth. "I already told ya."

Joe had the bar up into the door. He told Dave to give the end a tap with the ax.

"No ax," the guy shouted. "I don't want nobody chopping down my door."

Dave wheeled around to face him.

Cooper stepped in front of the guy. "Now look," he said. "I'd appreciate it if you'd keep quiet and stay out of the way."

The man's face went red, but he didn't move. Cooper took him by the arm and led him aside. A number of tenants were watching. Billy Cooper is a lanky, soft-spoken guy with a shock of brown hair he's always brushing out of his eyes. He's studying hard now for his Captain's exam, still months away, and he looks more like a schoolteacher than like anybody's idea of a fireman. "Now I'd appreciate it if you'd keep out of the way," he told the manager quietly.

Dave tapped the bar and the doors sprang open.

The girl was huddled in the far corner. She looked up at us wide-eyed, wiped a tear from her eye and then burst out crying. Joe comforted her and Cooper got the information he needed from her and from the manager; then one of the building residents carried the girl across the street to where she lived.

We hurried out to the Rescue truck. It was freezing cold and there was a full moon hanging just above the horizon. The manager followed us out. "I'm gonna find out your name," he called to Cooper. "You hear? You're gonna be sorry you ever met me."

"The name's Cooper," said Billy, climbing up into the cab. "You spell that with a C."

We returned to the firehouse. Cooper went to write up his report while the rest of us headed up to the third-floor kitchen for coffee. Jack Dillon, the Lieutenant of Engine One, and Jerry Martin, of the Aerial Tower, were shooting the breeze at the long table, and we sat down beside them. The wind rattled the windows and I cupped my hands around the coffee mug for the warmth.

Eddie Kilroy of Engine One tiptoed into the room with one finger to his lips cautioning silence. He had a broom tucked under his arm, and was trying to sneak up behind Dave, but Dave spun around and froze Kilroy in his tracks. Dave is six foot three, strong as a horse and quick for a big man. Ten years ago he was a basketball star at Rindge Tech, two blocks up Broadway. Now the golden locks are thinning out a bit and he's just touchy enough about going bald to make it worth Eddie's while. Kilroy is a feisty little Irishman with a mop of curly hair and the makings of a king-size ulcer. If you bounced him a basketball he'd shake it to see what was in it for him. Everything's grist for his mill. Easily the biggest slacker of us all when it comes to housework, he started sweeping around Dave as if there were no tomorrow.

"Well now," said Jerry Martin, "it's about time somebody's takin' an interest in the welfare of the house here."

"Well I couldn't agree mar, Mr. Mar-tin," said Kilroy, putting on the brogue. "It's a damn shame it is they let him in a place of eatin' with a condition such as that. Hair falling all over, littering up the floors. It's not proper."

Davey forced a smile and took a swig of his coffee. Kilroy peered at the back of Dave's head and pretended to be counting on his fingers. "Hell, Swanson," he finally announced, tossing the broom aside, "you still have almost fifty hairs. You've got another month at least."

Jack Dillon chuckled to himself and ran his fingers through his fine silver-white hair. Jack has more than twenty years on the Department and a wealth of stories tucked under his belt, and he isn't the kind to hoard them.

"Didja ever hear of Franny Boyle, Dave?" he said. "Franny Boyle from River Street."

"His brother was George Boyle the cop," said Jerry Martin, who goes back a bit himself. "His old man was on the MTA."

"That's the one," said Jack Dillon. "Well, Franny's hair started falling out in forty-six. I remember the year because Franny'd just been discharged from the Army. He had a good job as a warehouseman over in Somerville and some money from his old lady. John Crowley, who was going to law school nights at the time, convinced Franny over a couple of beers at Rafferty's that he'd be bald in a week. Next morning Franny draws the money from the bank and disappears. When he come back a month later he was drunk, broke and out of a job, but he had as good a head of hair as ever. And you know, he never did lose his hair after that and he never told us where he'd been. Not even Crowley knew, though of course he always said he did."

"I never seen a guy always had his way as John Crowley did," said Jerry Martin.

"Oh, yeah," said Jack Dillon. "I remember one time I

run into him on the street just after he voted against a Department raise. I says to him, 'Ya sonofabitch, John. Ya haven't changed a bit.' He says to me, 'Jackie, when we was kids together at St. Anthony's didja ever think you'd be calling the mayor of the city a sonofabitch to his face?' "

"He was a good mayor," said Joe. "Too bad he ever died."

"He was a bum like all of them from River Street," said Jerry Martin. "But that George Boyle was the real bastard. What ever become of him?"

Jack Dillon raked his fingers through his hair, smiled and leaned forward in his seat. All of a sudden the loudspeaker intruded: "*Attention. Rescue going out.*"

Billy Stone, Dave, Joe and I sprang to our feet. We slid one set of poles to the second floor, then another set to the apparatus floor. "Kinnaird Street," shouted John Riordan from the patrol desk as we ran for the truck. Dave steered us out onto Broadway.

"What's it for?" Joe yelled up to Cooper over the roar of the engine. Cooper looked back and shrugged his shoulders. He said it was a police call. Sometimes people first call the police and the police call us; other times people call us directly. But the cops try to make it to all fires and emergencies whenever they aren't tied up.

We pulled up in front of a single-family house. The cops were just getting out of their car. We rang the doorbell and stood there, our breath showing in the cold. The porch light came on and a woman opened the door partway. She was a tall, stern-looking woman, about sixty, with thin lips and narrow eyes, and she stared out at us with that special cold mixture of civility and contempt that I remembered from my days as a salesman.

"What's the problem?" Cooper asked her.

"He's upstairs," she said.

As soon as we stepped inside we knew what it was.

There's no mistaking that smell. When we opened the bedroom door at the top of the stairs the stench forced us to cover our mouths and noses. The old man lay there on the bed with his mouth wide open. He was stiff as a board. Cooper went up to him and pulled down the covers. The bed was soiled. I thought to myself, at least it's winter and he's not crawling with bugs.

Cooper told the cops there was nothing we could do and they went out to call for the medical examiner.

There was no light downstairs in the hallway. The living room was dark and cluttered with chairs under gray plastic covers. Cooper and I found the woman in the kitchen, leaning against the stove with her arms at her sides, her head lowered. A kettle of water was heating and a single place had been set at the table for tea.

"There's nothing we can do, ma'am," said Cooper.

The woman nodded without looking up. "I kinda figured that," she said.

"I'll have to have some information," he said, pulling out his note pad.

The woman shrugged. "I don't care," she said.

"Can I have his name and age?"

"William McCarty. Sixty-three."

"And your name, ma'am?"

"Mary."

"And your last name?"

"McCarty."

"He was your husband?"

"Yes."

"When was the last time you saw him, Mrs. McCarty?"

"What's the difference?" she said.

"It's for the report," said Cooper.

"Three days ago," said the woman.

"Three days ago?" said Billy, glancing at me. "Didn't you think something was wrong?"

"Why should I have?" said the woman. "We don't talk."

"But you lived together?"

"He pays me the rent; I cook him his meals. That's the agreement," she said, raising her head.

"Didn't he come down for his meals?" asked Cooper.

"He comes down most of the time. But sometimes he doesn't come down. Sometimes he goes a couple of days without touching a bite." She crossed her arms and looked past us. Cooper closed his notebook and put it in his back pocket. The kettle started to whistle. Two policemen were waiting at the door to question her. "A funny man," she said to herself, shaking her head. She didn't move to turn off the kettle. It was still whistling when we stepped out of the house onto the street and breathed the fresh clean air.

On the way back to quarters Finnegan said it must have been the stink that had finally gotten to her, that if it wasn't for the stink she wouldn't have lifted a finger. Billy Stone, who was the youngest of us and the quietest, said that maybe she had found him up there when he was still alive and left him to die. "Oh, yeah, she knew," said Joe. "That broad knew."

"Maybe she just didn't care," I said.

"It comes out to the same thing," Joe said.

"Still," said Billy, "she might have saved him."

Who could know?

I thought of the old people in rooms alone who would have died if somebody hadn't called up and said there's this old lady, I haven't seen this old lady, she lives in our apartment and I haven't seen her for two days and I usually do see her. And then going down and busting in the door and jumping back as a starving cat rushes out and then finding the old lady on the floor in her own mess staring up at you unable to move or talk but making that sound they make at the throat when they want water.

I thought of the old ones who *had* died, especially in the

summer when the stench makes you reel as you open the door, because they keep their places shut tight, old people; but they can't keep them shut quite tight enough to keep out the buzzing flies and the squirming maggots.

It's awful to think of them dying alone, but sometimes it seems worse the other way. Those calls come in at six, six-thirty in the morning; old people go in the night. One of them wakes up and gets out of bed and the other one doesn't because he's been dead three hours. They've lived together all their adult lives and now the one that's still alive waves you over to a corner and whispers in your ear for you to take away the one that's dead as quickly as possible so she can make the bed. She's made the bed first thing in the morning every morning for fifty years and she's got to get on with it. And it really doesn't matter, you tell yourself, because they stopped living a long time ago.

Back at the house John Riordan was on the pay phone. I wanted to call Bev; he cut short his conversation when he saw that I was waiting. "Whadja have?" he asked me, hanging up.

"Some guy passed away," I said. "His wife couldn't smell too well."

I dropped a dime into the phone. My three-year-old son had kept us up the night before with a cough, and he'd been running a temperature all day. Suddenly the bells hit. I hesitated for an instant, hearing the phone ringing in my hand, then hung up.

The floor was alive with bells and commotion and men shouting to each other as they slid the poles and ran for the apparatus. As I ran I counted the bells: six and then nine and then two, Box 692, Linnean Street. Everything was going. The overhead doors lifted open. Petey Hendricks, the Deputy's driver, slipped and fell and scrambled to his feet while the Deputy climbed into the car. Billy Stone unplugged the battery charger, Dave started the engine and Joe came running with a slice of bread folded neatly between his teeth.

"All set," I yelled up to Cooper. We pulled out of the house for the run up Mass. Ave. Behind us came the Deputy, the pumper, the hose wagon and, swinging out last of all, the long Aerial Tower.

Joe Finnegan wolfed down his sandwich while Billy Stone and I got into our coats. The rubber hoses on the masks were swaying and the stretchers rattling with the motion of the truck. The voice of Joe Flynn, the Fire Alarm operator back at the house, came over the radio: *"Fire Alarm broadcasting an alarm of fire box six-nine-two opposite one hundred Linnean Street."* It's the job of the Fire Alarm operators to monitor all the boxes and "stills" coming in and to dispatch the necessary companies and coordinate all communications. They aren't firemen. They have to take special exams to qualify for their job, which demands technical training to handle the sophisticated electronic equipment that's crammed into their headquarters.

Up forward Cooper blasted the air horn, sending the cars ahead of us scurrying into the right-hand lane as we flew by them. One woman was trying to outrun us. The air horn trumpeted angrily, but she wouldn't pull over. Dave was breathing right down her neck. He slammed the brakes hard as the light signal ahead turned yellow. The woman ran the light. The air brakes hissed as the Rescue slowed, cornered and roared up Linnean Street. We could see the flashing lights of other apparatus ahead.

The radio crackled: *"Engine Four to Fire Alarm."*

"Engine Four," said the Fire Alarm operator.

"Engine Four by the box: nothing showing."

The Rescue rolled to a stop. We climbed down and walked ahead with the crew from Engine One. Fire apparatus jammed the narrow residential street, filling it with flickering lights and the hum of motors and the acrid smell of diesel fumes. It looked like a convention: besides us, the

Deputy, the Aerial Tower and Engine Company One, there were Engine Company Four, Engine Eight and Ladder Four. While the Deputy Chief and Lieutenant Boyle of Engine Four inspected the box, the rest of us roamed around in boisterous packs calling out to friends from other companies. Faces peered out at us through parted curtains. If this had been a fire and not a false alarm, we and they would have been glad for the quick response of every piece of apparatus. Now there was an element of the ridiculous about it all. Tramping about in our boots and helmets, carrying axes and bars, we must have looked like drunken revelers crashing a quiet party.

Lieutenant Boyle rewound and locked the box and the Deputy ordered the "All Out." We trooped back to the Rescue. Six minutes after the box had been struck we were headed back to quarters.

Last year we responded to more than a thousand false alarms. Some of them were "stills"—calls that come in over the phone—but most of them were box alarms. The overwhelming majority of them were malicious. They're an unavoidable part of the job. Last year in this country fires claimed twelve thousand lives and caused hundreds of thousands of painful, disabling, disfiguring injuries. They caused nearly three billion dollars' worth of property damage. Fire is fast and deadly, and as long as time remains the single most important factor in fighting fire, the alarm box will be the primary outpost in our defense against it.

It's easy to turn in a false alarm, because pulling a box has to be made as simple as possible. And as long as there are kids who don't understand the seriousness of turning in an alarm and adults who just plain don't care, we will have to live with false alarms. In this city, every time somebody turns one in, three Engine Companies, two Ladder Companies and the Rescue will respond. The lives of firemen and others along the route will be endangered, and the well-being of every resident will be placed in jeopardy by the resulting shortage of coverage. Because every time a box strikes we have to hit the apparatus, don our gear and prepare ourselves for the fact that in half a minute we may be working blind, groping our way toward the flames. No one takes the prospect of a fire lightly; last year more than 60 percent of our Department were injured in the line of duty.

I telephoned Bev as soon as we returned to the house. Larry was asleep, his temperature almost normal, but Helen, our twelve-year-old niece, for whom we're legal guardians, was coming down with a cold. I told Bev to put her in our room for the night. She said it was just a sniffle and it didn't matter if they slept in the same room. I told her not to take any chances. "What kind of a night are you having, Larry?" she asked.

"Not bad," I said.

Dave collared me in the kitchen. He wanted me to watch a TV program about conservation in New Hampshire; part of it had been shot near the land where he hoped to build a summer house. The summer before last he had actually started work on the foundation. He was always bringing in do-it-yourself books and articles on home construction to show us how easy it was. Sometimes late at night he'd be up in the lounge alone, his big frame bent over a sheet of paper on which he was drawing plans. If you went in he'd show them to you, but after a while he'd crumple the paper into a ball and toss it into the basket.

Kilroy, Martin, Kelley and Harrington firmly occupied the TV room. They were engrossed in some spy picture. "For Chrissake," said Kilroy as Dave made his move for the set, "we've just uncovered a Commie plot to take back the whole of goddamn Alaska. We can't be bothering now with some crummy half-acre in New Hampshire."

In the patrol room down on the apparatus floor Billy Stone, who had the eight-thirty-to-twelve watch, was tuned to the same picture, but he let Dave change channels. There was an aerial shot of weekend traffic at a cloverleaf exchange on Interstate 95. "That's the exit before the one I take," said Dave, standing beside the set like a teacher at the blackboard. The part filmed near his land consisted of an interview with a spokesman for the local tourist industry and a state senator, who wanted progress with caution. Dave switched it off even before it had ended. "You can't get any idea of what it's like from that," he said. Then he offered to buy us tonics.

Between ten and eleven o'clock we had two false alarms back to back, long runs down to East Cambridge. Afterward we watched the late news in the patrol room and the start of "The Tonight Show." A stand-up comic from New York was on, telling the usual jokes about subways, muggings in Central Park, pollution and Mayor Lindsay. The guy wasn't too good.

"Why does everybody from New York talk funny?" said Joe. "Whadda they all got, colds down there?"

"You ever hear yourself talk?" said Dave.

"Sure," said Joe. "I've taped myself lots of times."

"Howdja sound?"

"I don't talk funny," Joe said.

"Yeah, well I thought I didn't talk funny either," said Dave. "Then one night I'm watching Johnny Carson and he's out in the audience talking to some guy. I'm saying to myself, 'Geez, that guy talks funny, where the hell's he from?' So then Carson asks him where he's from. 'Bahston,' says the guy."

"That reminds me of this friend of mine," I said. "He was coming up from New York to visit and he called for directions. So I told him to take Route Two to the rotary. He said, 'Route Two to the what?' 'To the rotary,' I said. 'Rotary,' he said, 'what's a rotary?' 'What's a rotary?' I said. 'It's, you know, a place where two or more roads meet and go around.' 'Oh,' he said, 'you mean a soycle.' "

Everybody laughed. Carson cut for a commercial and the phone to Fire Alarm rang. Billy Stone cradled it and started writing quickly. We were up on our feet before he clicked on the microphone. *"Rescue going out,"* he announced over the loudspeaker, and then he sounded the Rescue's signal—one short ring and one long. "Central Square," he told us as we hustled to the truck. "Somebody cut up badly." Inside the Rescue we filled our pockets with gauze and bandages.

Three minutes later we pulled up in front of the post office. It was well after midnight and the steps were deserted. A man about fifty, wearing an army jacket, was leaning against a parked car, his hands shielding his left eye. There was a kid beside him. "I found him lying on the sidewalk," the kid said. The man's shirt and jacket were unbuttoned, his undershirt ripped, his face and neck streaked with blood.

"How are you?" said Cooper.

"Fine," said the man, reeling slightly. He was obviously good and drunk. "How're *you?*"

"I'm asking you how you feel," Cooper said.

"Fine. How're *you* feeling?"

"Let me see your eye," I said, and as I reached out he swung at me with his elbow. Dave grabbed him from behind.

"Let go of me, you cocksuckers," he hollered.

"We're trying to help you," Dave told him.

"I don't need ya fuckin' help."

"Let me see it," I said and I pulled his hand away. The eye was smothered in blood. He flinched as I lifted the eyelid. The wound ran along the underside of the brow. Whatever had hit him had missed his eyeball by a quarter of an inch. He was lucky. I had seen a lot worse, boxing in the Army.

Cooper examined his chest and sides. When he touched his ribs the guy cried out.

"Look, we're taking you to the hospital," said Cooper.

"No you're not," said the drunk.

"How old are you?"

"Twenty."

"Why don't you cut the shit?"

"Why don't you go fuck yourself?"

"How old are you?"

"Twenty-two," said the drunk, grinning.

"Get the chair, Billy," said Cooper.

Joe and I dressed his eye, watching him closely in case he felt like taking another poke at us. As we lifted him into the chair he started hollering and swinging. We strapped down his arms.

"Watch his legs, Billy," shouted Cooper as the drunk started kicking.

All of a sudden the guy's head fell forward and his body went limp. We wheeled him over to the Rescue, lifted him aboard and pulled out for Cambridge Hospital. I raised his head up. He resisted as I tried to open his good eye. When I opened it the pupil dilated immediately. I let his head fall forward.

"Son of a bitch," said Joe.

"He deserves an academy award," I said. "He's a real beaut."

"He's a son of a bitch," said Joe.

Joe hates drunk runs. We all do. You can't ride the Rescue in this city without getting fed up with them. When drunks aren't causing serious automobile accidents, or falling down stairs or out of windows or off barstools, or getting beat up in barroom brawls, they're faking injuries. When they act tough, get nasty or vomit all over you, it's hard to remember that alcoholism is a disease and that you're there to treat them for their medical injuries.

It seems hopeless. There are bars the length of Mass. Ave., literally from one end of this city to the other. In the gray light of morning you can see the drunks lining up outside the bars in Central Square waiting for them to open. I once read that the Greater Boston area, which includes Cambridge, ranks second among American cities in its rate of alcoholism.

We rolled this drunk into the Emergency Room and deposited him on one of the beds. He lay there with his arms dangling over the side, faking unconsciousness. "Well, what did you bring me now, Larry?" said Mary Lane, a veteran Emergency Room nurse. When we were kids Mary was the big sister of a friend of mine and we know more about each other than we sometimes like to admit.

"He's a turkey, Mary," I said. "You're going to have to sweep the feathers out."

She winked at me and grabbed the smelling salts and passed them under the drunk's nose. Right away the guy opened his good eye wide, shook himself like a dog stepping out of water and started to moan. It wasn't a bad imitation of the Hollywood version of coming to.

"Come on," said Mary, shaking him. "Wake up. Let's go."

"Hey, cut it out," he cried. "You better treat me right. I'm a veteran."

"Is that right," said Mary. "Veteran of what?"

"Whaddaya think of what?" said the drunk. "Of the Second World War, that's what. What else?"

"Oh, so you're a veteran?"

"That's right, doll. I helped win the war."

"That's something I never understood," said Mary.

"Wuz-zat, doll?"

"How we ever won with guys like you."

"Ya never seen the other side, darling. They was worse'n we was."

Mary laughed. "OK, sweetheart," she said. "Let's see

what they done to you now." And like a good boy the drunk turned his head so she could unwrap the bandage.

The guys who drive the private ambulance had just arrived with pizza and submarine sandwiches for the night staff. We squeezed into the tiny staff lounge with two of the younger nurses, an intern and Mary Lane, poured ourselves coffee and managed to scrounge up some cookies.

We were all shooting the breeze when the Chief Resident appeared in the doorway. A young blond woman, thin but nicely built, in her print dress and sandals she looked like any young woman you might see in Cambridge. The stethoscope dangling from her neck was the only difference. She had a habit of never quite looking at you, and she wouldn't speak to any of us on the Rescue unless it was absolutely necessary. Now she rubbed her eyes and the back of her head so that we could all see she had just been awakened. She didn't want any pizza. She didn't want a part of anybody's sandwich. She poured herself a cup of coffee and turned to leave. The drunk, his face still bloody and unwashed, his shirt still unbuttoned, stood tottering in the doorway, blocking her way.

"What do you want?" she said.

"I want the nurse to have her pizza," he said. "I said she could have pizza and I want her to have pizza."

"My name is Doctor Stoneham," said the doctor, thrusting out her hand. "I'm the Chief Resident. Who are you?"

"Who are *you?*" said the drunk, looking uncertainly at the outstretched hand.

"I'm in charge here. I'm from upstairs."

"Oh," said the drunk, pumping her hand. "I'm in charge downstairs. Glad to meetcha, Doc."

"I am asking you who you are, sir."

"Who I am? I don't know who I am."

"What's your name?"

"Dumbnutswitz," said the drunk.

"Mr. Dumbnutswitz?" said the doctor.

"That's right," said the drunk leering. "Say, Doc, have ya got a smoke?"

"No, I haven't," she said, squeezing out past him.

"Anybody here got a cigarette?" he said. "Give me a cigarette, willya, pal?" he said to the intern.

The intern handed him a cigarette.

"Anybody got a light?"

I struck a match for him.

"Hey," said Mary Lane. "I got my pizza. Now you get back where you belong." The drunk stood for a moment looking foggily at her and swaying on his thin legs, and then blew one large perfect round O. He watched it hang in midair, then turned and vanished.

"He's not bad, that drunk you brought me," said Mary Lane. "You brought me a live one for once."

The younger nurses giggled.

"You can have him," said Joe, his mouth full of cookies, the walkie-talkie strapped to his shoulder.

"He's kinda cute," said Mary.

"Jee-sus," said Dave.

"He wasn't so cute when we picked him up," said Cooper.

"I don't care," said Mary. "At least he's not sad. The worst ones are the sad ones."

"The worst drunk," said Joe, munching on another cookie, "is the one that takes a poke at you."

"You guys are getting jumpy," said Mary. "The bells are getting to you."

"Mary," I said, "did I ever tell you about the drunk we had one time? This guy was the real thing—five-day stubble, the bloodshot eyes, the whole bit—and he was feeling his oats so we had to strap him in good. There was this new young intern on duty. He was the kind that if you brought in some guy who'd been hit by a truck asks what kind of truck. He talked in this high-pitched voice. He says, 'Why do you have that man strapped in like that?' I say, 'To protect him and to protect us.' He says, 'I'm here to help that man. I took an oath. You unstrap that man.' 'OK,' I say, unstrapping him, 'terrific. He's all yours.' The intern, he leans over the drunk. 'You're all right now, sir,' he says. 'I'm here to help you.' This drunk, his head's rocking like a boat and he's squinting real hard through one eye, taking as good a bearing on the intern's head as he can under the circumstances. Then POW. The drunk must have missed him by two feet, but you wouldn't know it by the way the kid was screaming. 'Help! Grab him! He's gonna kill me!' "

Mary smiled. "I still say the sad ones are the worst," she said.

"I don't know," I said.

"No, Lar," she said with a flat shake of her head. Her eyes bore into mine. "The sad ones are the worst." There were deep circles under her eyes and she had put on a lot of weight over the years, but beneath the extra layer of flesh I could still see clearly the face of the girl I had known as a kid.

We drove slowly back to the house. The moon had climbed to the top of the sky, the night had turned colder and a strong wind had come up.

The house was quiet. Just about everyone else was asleep. Cooper and Billy Stone went up to bed. Dave, Joe and I turned on the late movie, *The Amazon Creature*. It was about this man-size, birdlike reptile that was preying on the Indian women in the villages along the river. Joe, who among his many other specialties is an Amazon River buff, hit us with a few facts during the commercials: that the Amazon basin alone manufactures 15 percent of the world's oxygen, that the island in the mouth of the Amazon is bigger than

Belgium, that the waters of the Amazon muddy the ocean a hundred miles out at sea. Dave said Joe was beginning to sound like the witch doctor who had just captured the lady anthropologist and the white hunter. Dave said Joe even looked like the witch doctor. Joe didn't see it. The fact was that with his dark narrow eyes Joe did resemble him a bit. The Creature turned out to be a disguise worn by the witch doctor to scare off outsiders and maintain the loyalty of his people through terror. The witch doctor had even spent some time pursuing his doctorate at Harvard.

"I think I remember seeing him around," said Dave.

"Shut up," said Joe as the witch doctor handed the lady anthropologist a vial of the secret Indian drug she had been seeking.

"We gotta," she said.

"We gotta?" said Dave. "We gotta what? We gotta the drug? What, is this broad speaking Italian now, Larry?"

"She said 'obrigada,' " said Joe, his eyes glued to the set. "That's Portuguese for 'thank you.' You hear that around all the time."

"Is that right?" Dave said.

Joe didn't have time to answer: a burning arrow thwacked into the witch doctor's tepee. A thousand hostile Indians launched their attack on the village and the hunter and the lady anthropologist fled into the jungle. I dozed off.

The bells brought me back. We hit the poles. Box 625—Inman Square. Engine One was going, too. The Aerial Company and the Deputy were already on the apparatus floor, where they had to stand by in case there was a second alarm.

"Where's Stone?" shouted Cooper. The Engine Company started out. We rolled out onto the apron. Billy came running and tumbled red-faced into the rear of the Rescue.

"All set," Joe yelled, and the diesel roared as we chased the Engine Company up Cambridge Street.

Billy Stone rubbed his hands together vigorously. "They're asleep," he said. "I couldn't grab the pole. I hit like a ton of bricks." His hair was tousled, his eyes half-closed. He looked angry and confused at being awakened.

"Forget it," said Joe.

"This has never happened before," said Billy, slapping his hands together. "I hope we don't catch a fire."

"Nah," said Joe. "It'll be false. You know," he said, turning to me, "artistically speaking I'd hafta say that picture stunk. But they certainly did have some spectacular shots of the river at the end there."

"I fell asleep," I said.

"Larry, that river's sixty miles wide at the mouth. Imagine that. All the way from here to New Hampshire."

We were bouncing hard over the potholes, the air horn blasting as we ran the red lights.

"Fire Alarm announces an alarm of fire box six-two-five, opposite nine-eight-oh Cambridge Street."

"Message received," answered Deputy Chief Henders, the one responsible for the downtown sector.

"Fire Alarm to Car Two."

"Car Two," responded the downtown Deputy.

"Deputy, we're getting calls for an explosion at Tad's Hardware, one-oh-eight-eight Cambridge Street."

"Oh brother," said Joe. He quickly clipped his coat closed and leaned back, inserting his arms into the straps of the tank and jerking it onto his back. Billy Stone was trying to pull on his boot. He had the wrong foot. He shouted for his other boot. Joe threw it at him as we streaked through Inman Square.

"Plenty of smoke up ahead," yelled Cooper. The night was filled now with the screaming of sirens, and the whole length of the street ahead of us pulsated with flashing lights.

"Engine Three to Fire Alarm."

"Engine Three."

"Fire showing."

At Prospect Street Dave braked hard for a red light, sending Joe, Billy and me toppling into one another like bowling pins. "For Chrissake," Joe shouted. "Hey, Swanson, why the Christ can't you watch it?"

Dave just hunched down tighter over the wheel and Joe cut himself off, realizing suddenly where we had just passed. It seemed like only yesterday that Ladder One, speeding en route to a fire, had been struck there by a trailer truck. We lost two men, one killed, the other paralyzed for life from the waist down. They had both been good friends to all of us, but the man who died had been Dave's very best friend and it was Dave who had reached the wreck first and carried him out. Now Joe turned and threw open the rear doors. A passenger car trailed close behind us. "Get back," Joe shouted, shaking his fist at the driver. "You get back three hundred feet."

We pulled to a stop behind Engine One. Engine Companies Five and Three were up ahead laying lines. The hardware store, a single-story frame structure on the corner of Cambridge and Elm, was fully involved. The windows had blown, and inside the paint cans were exploding like artillery. Fire rolled out of the roof, sending smoke and burning cinders soaring into the sky. There was a five-story brick apartment building on the left. Flames were lapping up the side of it and the wind was bringing more fire down that way. The downtown Deputy was waving his arms at us. "Get 'em out of there," he shouted.

We ran into the building. People were streaming down the stairway carrying children in their arms. The children were crying and the people shouted at each other in Portuguese and broken English, fear and confusion all over their faces. Joe, Billy and I fought our way up against this human current, banging on doors as we climbed. On the third floor an enormous fat woman blocked our way. She pounded hysterically on the wall, her mammoth breasts surging and

flopping under her nightgown. "Ai," she screamed uncom-
prehendingly at us. "Ai. Ai." We pushed past her to the next
floor, where a tiny man, dressed in a dark suit and cradling a
small dog in his arms, gave us a half-salute and stepped back
to let us pass.

A thin haze of smoke showed on the top landing. Three
of the doors were open, the apartments clear. Joe busted in
the fourth door with the halligan: the apartment was empty.
We started back down.

On the floor below an entire family, men, women and
children, were racing around their apartment like rabbits,
grabbing things and throwing them into huge steamer
trunks. Through the window we could see the orange glow of
the fire and hear the explosions. There was no telling how far
the fire had extended, or what unknown explosives remained
in storage. Joe and I grabbed one of the men by the arms. His
relatives halted in their tracks.

"Never mind your stuff," said Joe. "Hurry up."

The man smiled humbly. "I have may-nee sings," he
said, turning his palms out. "Why hurry?"

"Forget your things," Joe told him. "There's no time."

"No time?"

"No. No time. Understand?"

"Si. OK." The man turned and shouted rapidly in Portuguese. They were all out and down the stairs in a flash.

The stairway was empty and silent now. With all the
doors thrown open and the lights blazing the building looked
like an abandoned ship. We still had to clear the third floor,
and in the back room of one apartment we found an old man
sitting on top of a suitcase. He wore suspenders and a hat,
and he was panting heavily and holding a hand to his chest.
The suitcase was stuffed to overflowing and he had been
unable to close it. "Come on, Pop," said Billy Stone, helping
the old man to his feet. He led him out.

Joe and I cleared the last apartments on the floor, then

met Dave and Cooper coming up the stairs. "All clear?" Cooper said.

"All clear," said Joe.

When we came out of the building we saw that the second alarm had been struck. The Aerial Tower's bucket rose high above the street like a giant steel monster. The sight of that piece filled me with dread: we were in for a long, cold, hard, wet night. Engine Seven was hooked up into the Tower and both the fixed aerial guns were going full blast. All you could see up there against the spotlights was the dark outline of the two men in the bucket directing the streams of water down into the fire and smoke. You couldn't look long, because the wind blew the freezing spray back into your face. Most of the fire seemed confined to the storage building directly behind the hardware store. Engine Companies Five and One, Ladder Companies Two and Three and other pieces were operating from around the corner. Additional second-alarm apparatus was coming up Cambridge Street past the police roadblocks.

The Deputy shouted for us to get a line up on the roof. He wanted us to shield the apartment building from the fire. Engine Three was jockeying a line to the front of the store. Cooper told Billy and me to help them while he took Dave and Joe to investigate the rear. Billy and I dropped our tanks and masks where we stood; the MSA tank-and-mask units aren't designed for heavy exertion like laying and jockeying lines. When you start breathing too hard through them you begin to suffocate. They're all right for rescue work, but the engine companies can't use them in their initial attack on a fire. Nothing I know compares with that for sheer exertion. A fire fighter burns up energy at a rate close to that of an Olympic athlete, but he is at the same time subject to incredible stress: the shock of being banged out of sleep by the bells, the anxiety of wondering whether the next minute will find him blind in smoke, the fear, once he's inside a burning building, of the dangers that might befall him.

This combination of exertion and stress exacts a steep price. Heart attacks are the occupational disease of firemen. We run a much higher risk of heart and lung disease than the general population does. I've read that a fireman's life expectancy is nine years less than that of the average workingman.

Ladder Three had thrown a ladder against the front of the store. The lead man, Art Sweeney, threw the nozzle over his shoulder and started up the ladder with the line running between his legs. Julie Kearns, Billy Stone and I flung loops of the heavy two-and-a-half-inch hose over our shoulders and followed him up. Sweeney laid the tip over the edge of the roof, stepped off the ladder and began hauling up on the line. The thick line flowed smoothly up the ladder.

There was heavy smoke on the roof and the flames shot up in front of us. The main body of the fire was farther back and to our right. Without the mask my eyes began to sting and water. We could hear the paint cans exploding below us as we clung to our position at the edge of the roof, Sweeney on the tip, me, Billy and Julie Kearns backing him up.

"Tell them to charge the line," shouted Sweeney over the roar of the fire. Julie hollered down to the street. The line bulged and slithered between our legs as it filled with water. Sweeney cracked open the tip, then slowly eased the lever forward. I leaned hard against him, cradling the thick line in my arms, and felt the back pressure as two hundred and fifty gallons of water sputtered, hissed, then roared through the tip. Sweeney fought to steady the stream. You can't play with a two-and-a-half-inch line. One of them once threw Davey and me into the air when a relief valve malfunctioned. Another time one drove us right out of a room and down a flight of stairs. Both times we held on to it.

If the thing ever got away from you it could spin around and kill someone. You have to fight it the whole time. You just don't have the flexibility and control that you have with a one-and-a-half. But whatever its drawbacks, the big line gives you a tremendous volume of water—two hundred and fifty gallons per minute of it. You can knock down a lot of fire with that.

"All right," shouted Sweeney above the roar of the water and the fire, "I'm gonna hit the exposure now." Reaching around him from behind, I helped him raise and pivot the nozzle a few degrees up and to the left. Sweeney played the line on the fire where it was shooting up out of the roof and lapping against the side of the apartment building. The fire retreated quickly.

"Let's hit it higher," Sweeney yelled. We raised the tip. "Take up on the line," he shouted. I passed the order behind me, and as Billy Stone and Julie Kearns pulled up some slack, Sweeney and I moved forward a step.

We played the line on the side of the apartment building, keeping the bricks good and wet. It was hard to see. The smoke was thick and swirling all around us in the shifting wind. My eyes burned and we were all coughing now. The freezing spray rained down on us. We could hear the crackling of the fire up close, and from time to time we glimpsed the flames raking the roof as the curtain of smoke opened and closed in front of us.

It was two A.M. but the night was lit up by the fire and the Aerial Tower's spotlights, shining down like fierce suns. There was another light, blood-red, suspended right overhead. A cold shiver passed through me. "What the hell is that?" I thought, nearly losing my balance. Then the smoke cleared for an instant and I saw it was the moon.

I thought I saw someone walking to the right of us. It was impossible to tell. The spotlights cast our shadows against the thick smoke as sharply as if it were a wall. I glimpsed someone again; then the smoke closed in around him and he disappeared. There was a rapid series of explosions below like an artillery barrage, and the curtain of smoke parted. There in front of us was the Deputy, his long white coat and high peaked helmet dazzling in the light of the fire, his arms spread out like great wings. He almost seemed to hover in midair; then he plunged.

"The Deputy's fallen!" I shouted.

"What?" yelled Sweeney. "Where?"

"Between the buildings. I'm going down." I shouted to Julie Kearns to take my place.

The ladder had iced up. Halfway down I slipped, lost my grip and fell to the sidewalk. I got up and ran into the narrow alley between the store and the apartment building. Flames were shooting out overhead. Billy Stone was right behind me and I shouted for him to get the orthopedic stretcher.

Deputy Henders lay unconscious, half submerged in water, in the narrow alleyway between the rear of the store and the warehouse. His face was cut and blood was spurting from the side of his mouth. Cooper was kneeling over him pressing a compress to his head.

"How bad?" I said.

"He's hurt bad," said Cooper.

The fire roared overhead and water was running down the walls and swirling around our boots. Joe and Billy Stone came running with the aluminum orthopedic stretcher. We separated the two halves, slid them under his body, connected them and rushed him out.

Steve Grady, the Deputy's aide, stood waiting by the Rescue, his face ashen. He tried to force his way past us to the Deputy. "Get back, Steve," said Cooper. "Sit in the truck." Grady climbed into the Rescue. Our faces red in the flashing lights of all the apparatus in the street, we lifted the Deputy aboard and laid him on the stretcher bench.

Henders' face bulged grotesquely. Blood was pouring from his mouth and his breathing was heavy. I parted his lips. His teeth had gone down through the jaw. His mouth was full of blood; he was in danger of choking.

"Give me the aspirator," I said.

I inserted the tube in his mouth and suctioned out some of the blood. He started to come around. He rolled his head from side to side, moaning incoherently, then heaved himself up against the straps. Joe pinned down his shoulders. "You'll be all right, Deputy," he said. "We're almost there."

"Let me up," Henders groaned. His face was completely blown up out of shape. You could hardly recognize him. He started mumbling deliriously, and every time he spoke blood oozed from his mouth. There was no telling if he was all broken up on the inside. There was no telling if he was going to die on us.

"How is he?" asked Grady. His voice sounded dry and very far away.

"He's all right, Steve," said Joe.

"He was just finally getting over that accident last year," said Grady, starting to get to his feet.

Joe shoved him back down. "Just sit still and be quiet," he said.

We took him into the Emergency Room. The doctors and nurses rushed in. We helped them take off his boots; they acted as though they didn't know us.

Steve Grady stood in the hallway just outside the door holding the Deputy's helmet in his hand.

"Come on," I said, taking him by the arm. "There's nothing we can do here." We went down to the room where just an hour and a half before we'd eaten submarines and kidded with the nurses. I handed Steve a cup of coffee. He held it in his hand without even looking at it.

"It happened so fast," he said. "How could it have happened? He only had a few months to go until he retired.

He was just telling me how he was finally starting to feel himself."

The cup in Steve's hand was shaking. I realized how he felt about the Deputy. Steve drove the Deputy's car, maintained radio communications between the Deputy and Fire Alarm and relayed the Deputy's orders to the various companies at the scene. He and the Deputy were together on the job day in and day out. They were close personal friends. I didn't know the Deputy that well. But the memory of his falling made me shudder, and I had to put my cup down on the counter.

They wheeled his bed out into the corridor. Tubes connected him to the plasma bottles hanging above his head. We all stood up, our eyes red and our faces black from the fire. Mary Lane poked her head into the room as they passed by. "Well, don't you all look a sight," she said. "What've you got going tonight, an orgy in an abandoned coal mine?"

"A two-alarm mercantile," said Cooper. "How is he?"

"A fractured cheekbone and a possible fracture of the jaw. They're taking him to X-ray. There's a nasty cut in his mouth, but he's talking rationally now."

"Thank God," said Steve Grady.

It was about a quarter after three when we returned to the fire. It looked just as bad as when we had left. We treated a couple of guys for smoke inhalation and minor burns and then took up positions with Engine Companies One and Four and helped jockey their lines. There were at least a dozen lines going altogether, each pouring roughly two to three hundred gallons per minute into the blaze. Water cascaded from the doorways, and one of the walls bulged from the pressure of the water inside. Deputy Simmons, the uptown Deputy, who had taken command of the fire, ordered us back well away from the building because of the danger of collapse. Not long before, over in Boston, the wall of a hotel had suddenly collapsed, killing nine fire fighters.

Now we watched carefully, and a few minutes later the roof on the rear storage building caved in. The flames mushroomed and shot into the air as the fresh oxygen fed the fire again.

My knee had begun to stiffen up from the fall I'd taken. Jockeying all that heavy hose wasn't helping it any. Outside fires may be less dangerous than inside ones, but they're a lot more work. And lugging a two-and-a-half-inch line, which is exhausting, backbreaking, boring, god-awful work under any conditions, in weather like this becomes brutal. The streets were slick with ice and it was difficult to keep our footing. The wind-blown spray stung our faces like pellets and the freezing water worked its way under our coats, soaking us from our T-shirts down to our socks. Tiny icicles formed on the fronts of our helmets. My ears burned from the cold, my nose was running and I couldn't stop my teeth from chattering. I slapped my shoulders, rubbed my hands, stamped my feet. There seemed no end to it. Shivering, I shut my eyes tight, trying to remember what it felt like to be warm and dry.

At four-thirty the fire was under control. The hardware store and the storage building behind it were lost, but the apartment building, though completely blackened by smoke on the near side, had been saved. One by one we were given breaks. We stood in the warm shelter of the pumpers, close beside their throbbing engines. Some of the neighborhood people opened their houses to us and even made hot soup and coffee. We stood in their hallways, stiff-armed like penguins in our frozen coats, too exhausted to say much more than thanks.

Around six o'clock, when the fire had been knocked down and only white smoke rose from the smoldering rubble, Deputy Simmons began releasing the second-alarm companies. We helped Engine Seven make up. Then he released us. The first streaks of gray showed in the east as we rode back to the house.

Just before seven we had a run up to North Cambridge. An elderly woman said she thought she had a gas leak. It was only the pilot to one of the burners on her kitchen stove, which had blown out during the night.

At seven-thirty we lined up for roll call at the rear of the Rescue, our squad standing behind the one that was coming on duty. Then I drove home.

It was a raw, steel-gray morning. Bev was keeping Helen home from school because of her cold; the baby was still asleep. I tumbled into bed. My knee ached and throbbed and my throat felt raw and smoky. When I closed my eyes I saw the Deputy with his arms outstretched. Twice I dreamed that I was falling through thin air.

=THREE

THE DOOR CLICKED OPEN. SQUINTING THROUGH
one eye I could see him sneaking alongside the bed. He
leaned over and examined me closely, disapproval written
all over his three-year-old face.

"Daddy," he whispered, "are you up?"

I started to snore.

"Daddy?"

I groaned and pressed my head deeper into the pillow.

"Are you up, Daddy?"

"No."

"Then are you dreamin' about me?"

I opened my eyes and stared past the faint freckles on
the soft tip of his nose and the unbelievably long lashes, into
his bright brown eyes. "Who are you?" I said, shutting my
eyes.

"*You* know."

"Oh, yeah. It's you."

"Yes. And you're dreamin' 'bout your promise, because
it's Christmas."

"What promise? Christmas is tomorrow; I don't remember any promises for today."

"Yes you do. Milk from a box."

"What else?"

"A jelly doughnut on a stool and talking to a doughnut lady."

"They're not worth talking to."

"Don't you even wanna doughnut?"

"I want to sleep. I think I have to sleep a little more."

"Why?"

"I have to. It's business."

"I'm business, too," he said.

I sat up and looked at him in amazement: all thirty-one and a quarter inches of him. He never quite seems real to me. The smirk on his face told me he knew he would have his way. But then he almost always does. I grabbed him by the hips and swung him screaming with laughter high above me. "OK," I shouted. "It's Dunkin Donuts time."

The bedroom door flew open. "Hey, you," said Bev, her hands on her hips. "What did I tell you?"

"What?" he said, looking as dignified as anybody possibly could when standing on somebody else's chest.

"Didn't I tell you to be as quiet as possible? Didn't I tell you to let your father sleep?"

He shook his head solemnly. "You told me see if you can wake him up before the whole day's over."

"Why, you little devil," said Bev, running straight for him. He squealed, flopped onto the bed and tried to burrow under my arm. She dragged him out by his feet and pretended to spank him while they both shrieked with laughter. Laughing, they looked exactly alike: the same quickness, the compact, energetic bodies. He squirmed free and ran from the room.

Bev smoothed down her hair and sat down on the edge of the bed. "Well, did you get enough sleep? You looked like a zombie when you came in."

"I felt like one. We had a Sunday night special: some

fruitcake walked from Harvard to Kendall Square pulling boxes. We were chasing false alarms all night. What time is it?"

"Almost noon," she said, brushing my hair back. "I thought you'd want to ge . up." Her fingers traced along the outline of my ear and down the side of my neck. Then she kissed me on the forehead.

"Can you smell it?" I said.

"Yes."

"It was just a junk fire. It wasn't much."

"You weren't hurt, were you? Larry?"

"I already told you. It wasn't anything."

"That's not what I meant."

"I thought you weren't doing a laundry today."

"I wasn't going to," she said, "but I did. You didn't hide the pants that well; I had to run them through twice. At least now they can't walk by themselves," she added.

I told her I would throw them out.

"Of course not," she said. "They're all right now. It doesn't matter. And now we've got the whole day."

She told me my brother Ray had called to say he might come by to drop off the kids' presents. Ray makes his living training horses up in the northern part of the state. He doesn't get to Cambridge too often and Bev thought we should really have a present ready for him if he did. Helen had been wanting to go Christmas shopping anyway, and Bev had a few last-minute things to take care of herself.

"You don't have to go in early, do you?"

"Nope, five-thirty."

"Good," she said, getting up. "We're going to have a fine day. I'll dress your boyfriend for your date."

"Bundle him up good," I said. "It was really cold last night."

Half an hour later my son and I were perched on stools at the Dunkin Donuts up in Somerville. It was a bleak, gray afternoon, with patches of old snow on the ground, and the place was packed with workmen lingering over their coffee. My son's attention was riveted on the waitress who was getting our doughnuts. She had freckles and carrot-colored hair that she wore in two braids. She put our doughnuts down in front of us. "What's your name?" he asked her.

The girl smiled. "Jeannie," she said, pointing to the nameplate pinned to her blouse. "See?"

"Jeannie's a nice name, Jeannie," he said.

"Thank you," said the girl. "What's your name?"

"My name is Philip," he said, leaning forward and blowing bubbles through his straw into his milk.

"Wow," she said. "That's a terrific name."

"I wanna change it," he said, twisting from side to side on his stool.

"To what?"

"Larry."

"Oh, don't do that," she said, winking at me. "Philip's a perfect name for you."

The girl left us for another customer. He spun around and we both cracked up. I don't know where he gets his stuff.

"Hey, don't spin so much," I said. "And drink your milk, Larry."

"Philip."

"Philip nothing," I said. "Larry's good enough for me, it's good enough for you."

We drove back to the house and waited outside for Helen and Bev. We live in a two-bedroom apartment in a two-family house on Harvard Street. One of the old Mac-Donald sisters, who live next door, was coming up the sidewalk with a shopping bag. She wished us a Merry Christmas but said she didn't think people had the same old holiday feeling, nowadays.

"I don't know if you read the *Globe* last night," she said, "but there was this piece by a man who said he was a woman's doctor. And I said to Mary I'd never read such stuff in my life. To put it out in bare print."

I mumbled something about changing times.

"Change is an understatement, Mister Ferazani. We seem to be in a Mixmaster. We don't know where we're at. I was just up at my sister-in-law's in Arlington. Poor thing, she already wasn't feeling well and now two young girls have moved into the apartment above her. They had just a foam-rubber mattress. Sally said where's the bed? 'Oh, we just sleep on the mattress,' they said. I don't understand these kids. They don't have no furniture. And these were beautiful, fairly intelligent girls. Can you imagine? And Sally wanted to give them a divan. Sally's one of these people whose heart is bigger than her body. Mary Moriarity, poor dear, was another."

She began to tell me the story of Mary Moriarity's wake, but mercifully Bev and Helen emerged from the house, and I was free to make my escape.

We drove up to the Fresh Pond shopping center, where "I'm dreaming of a white Christmas" was blaring over the loudspeakers and under the arcade a Santa Claus stood ringing a bell beside a charcoal fire. Bev took Larry and gave me the grocery list.

Helen and I headed for the Stop & Shop, but first she wanted to go into a women's store. It was the kind of place where the saleswomen jingle with jewelry and reek with perfume and when you get up close you find out they're caked with makeup. Helen had seen a bracelet there for Bev. She roamed thoughtfully among the counters and the racks of clothing, clutching her wallet in her hand. She's a serious, dark-haired child and I think she looks a lot older than twelve.

This was her first Christmas with us, and already there had been a tremendous change in her. At first she had been loud and sometimes openly hostile. She hadn't wanted to come to us. She was a tough kid and she had seen it all after her father, Bev's brother, had died. All she ever wanted from me was to know if the Rescue had been to the Roosevelt Towers project. That had been the last stop on the way down before her mother died, too.

She was quieter now, but it was hard to tell whether her silence was a wall she was hiding behind, or simply the first rays of a natural shyness showing through. We hadn't thought it would be easy. But we had wanted a daughter badly and we weren't sure we could have other children. Bev's sisters had agreed to divide up the four children; Helen was the oldest and we had wanted her the most.

She came toward me. "See anything?" I asked.

She shook her head.

"What about that bracelet?"

"It looks junky. I saw another one but I don't have enough."

"Have you got anything else for her?"

"No. Only what I made in school."

"What was that?"

"A clay animal. It's a mongoose, sort of."

"A what?"

"Don't you know what a mongoose is?"

"Sure I do. We have a couple of them on the Fire Department."

"No, you couldn't, because they live mainly in India," she said. "They're very brave and they kill rats and snakes. They look like a long cat, only I made mine too tall so he looked more like a skinny dog. Then his front legs got cracked in the kiln and when I was wrapping him the other night they both broke off. Now he looks like a kangaroo, only worse. It's awful. I don't even wanna give it to her."

"You could glue the legs on if you still have them."

"I do, but you need a special glue that takes overnight to dry."

"Zayre's would have it," I said. "Come on."

"What about the bracelet?" she said, following me out the door.

"Forget it. Bev would much rather have something you made."

"But it's kind of silly-looking. You can't even tell what animal family it belongs to. But the glaze is really beautiful. He's reddish brown and he's got this white spot on his back. He's so shiny you can see yourself in him. Do you think she'll like him, Uncle Larry?" It was a lot for Helen to say.

"I'm sure of it, sweetheart," I told her.

When we got home Bev put Larry down for his nap and I helped Helen with the gluing. One of Bev's sisters, Alice, dropped by to see if she could talk Bev into going to her mother's that evening. She didn't see why Bev had to be stuck home Christmas Eve just because I had to work. Bev could get a baby-sitter or bring the kids, and they would give her a ride over and back. But Bev said it was too late to get a sitter and she didn't want the kids up late anyhow. Besides, she said, she wouldn't feel right going out if I was working.

At four o'clock, before eating supper, I went into the bathroom to shower and shave and put on my dress uniform. I was combing my hair when a bottle of vitamin pills on the shelf beside the sink caught my eye. I took it with me when I went back to the kitchen, and asked Bev what the pills were doing out on the shelf. She said she'd been forgetting to take them, so now she left them out. She poured herself a cup of coffee and sat down opposite me at the table.

"Bev," I said, "I thought we talked about this."

She said yes but vitamin pills didn't count.

I told her they counted, that we shouldn't have any medications whatsoever out where Larry could get at them.

"But they're only vitamins, honey," she said. "Alice

leaves hers right out on the kitchen table and she's got smaller kids."

"He's not Alice's son. He's mine."

"He's my son, too, Larry, and I don't want us making a nervous wreck of him. I can't be over him all day long. And you can't always be over me."

I asked her if she had ever seen a poisoned child. I told her how some of them were a lot older than Larry. I told her how they didn't even cry, how they made a sound like a small animal and just looked at you and all the while the stuff was burning their insides out. I told her about this woman who left her vaginal douche out. When we took her son away the woman said to me she couldn't understand how it had happened. How the hell did she think it had happened?

Bev shut her eyes and started to cry. She got up and left the room. I ate my dinner alone.

But when I put on my jacket and my cap she came out into the hallway as always, and she called the kids. Helen had been in the middle of dressing Larry and he burst into the living room and pranced around the Christmas tree wearing his pants at half-mast.

"Pull up your pants," I said to him. "What's Santa Claus going to say if he sees you like that?"

He turned around slowly and smiled at us. "He'll say 'Ho ho ho,'" he said.

At the door Bev put her arms around my neck and kissed me. "Take care of yourself," she whispered.

It was dark outside and the ground was crusted with week-old snow. I warmed up the car and backed out of the driveway. Framed by the Christmas lights around the door, the three of them stood waving good-bye.

Suddenly I felt all choked up. A man is lucky enough to have a woman who loves him and a family he can love. All he wants out of life is to live long and well enough to enjoy them, and to see his children grown up. They're just a few people, but they're his whole world and they seem larger than life to him. Then he steps outside and sees them from a distance and they look small and vulnerable, and he realizes that all the love in the world can't protect them.

Jack Dillon had brought a ton of cold cuts down to the house, boiled ham, Genoa salami, provolone cheese, rolls, potato chips, even ice cream. Dave had brought in a couple of pies. The idea was we weren't going to touch any of the stuff until just before the Bruins game came on at eight-thirty.

Either the kitchen was the warmest place in the house or nobody trusted anybody else near the food. Everybody was up there. Even Cooper and Lieutenant Yates decided to lay aside the books for a while. Some carolers came around and we went to the windows and they sang for us. Then, a little after seven, Eddie Kilroy climbed up on a chair with a bottle of tonic in his hand and asked for silence. He got a solid round of boos. Finally John Riordan yelled, "Shaddup, let's hear the bum out."

"Thank you, John," Eddie began as the boos died down. "I appreciate those kind words. Gentlemen, in every corner of the world tonight when men's thoughts are of peace and goodwill, let those of us gathered here tonight in this firehouse endeavor to remember that if the good Lord had truly intended for us to be pure of mind and heart this evening he most certainly would not have scheduled our beloved Bruins against the Rangers in New York."

There was cheering and applause. Then Jerry Martin was on his feet waving for silence.

"I would just like the opportunity to say how much I hate the New York Rangers," said Jerry. "I hate every god-damn thing about them. I hate their uniforms and the way they skate out at the beginning of a match and I hated them before they got to be good. I don't even wanna talk about it. Thank you."

The guys gave Jerry a nice hand. The room was hazy with cigarette smoke. Finnegan was up now predicting the Bruins would win by at least two goals.

"It's gonna be a great evening," Joe shouted happily.

"Jesus Christ," said Kilroy, waving his hand in front of him and making his way to the window. "We've got enough smoke in here for a two-bagger." Suddenly he threw open the refrigerator door, grabbed a salami and cut sharply to his left, making a neat end run around the table smack into the arms of Dave Swanson. Grinning from ear to ear, Dave gave him a bone-crushing bear hug and lifted him off the ground.

"Oh puh-leese, Davey baby, don't stop," Kilroy lisped, fluttering his eyelids. Swanson dropped him like a hot potato.

"Hell," said Kilroy, straightening out his shirt, "with any kinda blocking I woulda made it. Besides," he said, holding up the salami, "I ain't used to running with Italian footballs. Say, Larry, didja hear about that Italian kid that intercepted a pass and run it back ninety yards against Holy Cross?"

"Yeah," I said. "The one that would have scored a touchdown if he hadn't kept stopping every ten yards to ask directions."

"I already told you that one, huh?"

"No, but it reminded me somewhat of the Italian racing driver that came in second at Indianapolis."

"I told ya that one, too, huh?"

"Kilroy," I said, "did you ever hear of the Irishman who told a new joke?"

"No."

"Neither did I."

Kilroy shook his head and clicked his tongue. "Larry,"

he said, "you really are crazy. The only other Guinea in the house besides you and the salami here is in the refrigerator, and he's a cheese. Here you are surrounded by at least a dozen true sons of Ireland and you're casting aspersions on the race. Do you know what's wrong with your entire perspective? It's wopped."

Everybody broke out laughing, including me. There's no winning against Kilroy and there's no use getting angry with him, since he's not the least bit prejudiced; sooner or later he ends up knocking everyone equally.

Jack Dillon started slicing up the cold cuts. A few of the guys wandered off to claim the best seats in the TV room. Then the ring of the telephone echoed through the house. An instant later: "*Attention. Rescue only. Rescue going out.*"

"Oh, what a shame," said Kilroy as we jumped up from the table.

"You just save us something to eat," yelled Dave. We could hear them laughing as we slid the poles.

The address was in North Cambridge, a report of a shooting. We turned down a narrow residential street, where the houses were all strung with Christmas lights, and pulled up in front of a darkened two-family house. The cops weren't there yet. We knocked. The door was open, but we hesitated, not knowing who had been shooting whom.

"Aw, what the hell," said Cooper, pushing in the door.

In the kitchen there was a young boy by the phone. He leapt to his feet and led us down the back stairs to the cellar. The air was sharp with the acrid smell of burned gunpowder. At the bottom of the steps was another boy, about fifteen, clutching his leg with both hands and twisting his head from side to side. He had a gaping hole in the back of his thigh.

A third boy stood against the wall sobbing over and over again, "I'm sorry. Oh, Jesus, I'm sorry." He had tied two tourniquets around his friend's leg, one above and one below the wound. It didn't make any sense; in the first place there wasn't enough bleeding to justify a tourniquet at all, but in any event why two tourniquets? People who live in cities seem to learn only the kind of first aid appropriate for the Rocky Mountains. They know how to apply a tourniquet for a victim bleeding to death in the wilderness tens of miles from the nearest doctor, but they don't know that direct pressure is the simplest, safest, surest way of stopping bleeding in a situation where emergency medical help is only minutes away. People in cities know how to suck out snake poison, make splints, improvise stretchers; but they don't understand the absolute danger of moving an auto-accident victim when trained emergency personnel will be arriving literally in seconds. Victims who might have had a chance for complete recovery are often permanently paralyzed only because their would-be rescuers didn't know enough to leave them alone.

Cooper cut off the tourniquets. The kid was breathing heavily. "How bad is it?" he asked, his voice choked with fear.

"Not so bad at all," said Dave.

I placed a sterile gauze over the wound. The flesh was neatly gouged out. It must have been a shotgun, I thought, to have made a hole like that. The four-by-four gauze barely covered the wound and I had to hold it in place while Cooper applied several others and then loosely wrapped a cravat around the leg. The boy's eyes were riveted to our every movement, his face drawn with fear and pain.

"What happened?" Dave asked him.

"Me and Jerry," he said, trying to catch his breath, "we were foolin' around. It just went off."

"Shotgun?" Dave said.

"Yeah. It's by the table," he said. In the center of the room was a Ping-Pong table piled with ice skates, hockey sticks and pads. The gun was on the floor.

"Do you think I'm gonna lose my leg?"

"No," said Dave. "It isn't that bad."

"Well, how bad is it?"

"It's not that bad."

"But, Jesus," he cried, "it hurts me so bad."

The cops had arrived and were questioning the other kid. Joe and Billy brought the stretcher down, and Cooper, Dave and I lifted the wounded boy gently as they slid it under him. He rolled his head from side to side and suddenly tried to sit up and grab at his wound. We strapped him down and Dave told him to lie still. I covered him with the blanket and he asked me if he'd ever walk again. I told him that he was a very lucky guy, that the wound didn't look that serious. He didn't believe me. He asked Dave if it was true. "Sure, son," Dave said. "You're gonna be fine."

"My father'll kill me," he said.

"No, he won't," said Dave. "You'll see."

A small crowd of neighbors had gathered out in the cold and as we emerged from the house the cops helped clear the way for us. One of the cops, Timmy Connors, was an old pal of mine; he used to hang around Kerry Corner with me during high school, and when we were just kids we used to play street-hockey games. The last time I'd seen him he was throwing up, sickened by the effort of giving a gravely ill, vomiting man mouth-to-mouth resuscitation.

"Lar, how are ya?" he said. "Hey, I know the kid's old man. How is he?"

"It looks like he might have a long haul," I said, climbing up onto the rear step of the Rescue.

"Well, have a nice Christmas," he said.

"Same to you, Timmy," I said.

We ran the boy down to Mt. Auburn Hospital and then headed right back to the house, where we hoped to get some of the sandwiches and catch some of the game. But we had just backed into the house when the phone rang. "*Attention.*

Rescue going out," shouted Ken Hovey from the patrol desk. "An automobile accident on Concord Ave. in front of the Sheraton Commander."

As we pulled out we could see the neon light of the hotel straight across the Common. Joe maneuvered the Rescue past a line of blocked traffic. A pickup truck had slammed into the side of a convertible; you could see the skid marks on the icy road. The cops were talking to the driver of the truck. The car's fender was crumpled and the doors thrown open. There were six people sitting inside: a blonde behind the wheel with a sailor beside her, and two sailors and two other women in the back seat. They were all sprinkled with glass. Cooper asked them if anybody was hurt.

"We're fine, officer," said the sailor beside the blonde. "When can we get outta this crate? My wife and me, we got little kids at home."

"We're just going to take you out one at a time and make sure you're all right," said Cooper.

"I don't feel so good," said the blonde. "I feel a little shaky."

"Oh, you're fine, honey," the sailor said, dabbing at his bloody lip with a handkerchief.

"Yeah, we all feel terrific," said one of the sailors in the back. The two girls smiled. Their hairdos were rumpled; one of them had a gaily wrapped bottle-size package on her lap.

Only the blonde needed the chair; the others were all walk-ons. One of the women limped slightly on her high heels; she bent down to take them off. Dave offered to carry her and she giggled when he picked her up. A bunch of cab drivers gathered on the sidewalk in front of the hotel and whistled as we helped the women up into the Rescue.

"Jesus," said Dave, climbing aboard, "now I know how it feels on the vice squad."

We closed the doors and pulled out for Mt. Auburn

Hospital. Billy Stone gave the blonde some oxygen. The others filled the two benches, and Dave and I stood up. The girl Dave had carried smiled at him; he smiled back.

"It's very cozy in here," she said.

"Yeah, well, you know," he said, "we try to keep it nice."

"Oh yes, it's very nice," she said. She wore a good-size diamond ring, and bits of glass sparkled in her hair.

"Where're you guys from?" asked the sailor whose lip was cut.

"Right here," I said.

"Which is like where, man?"

"Cambridge," I said. "It's just next door to the Charlestown Navy base."

"Cambridge," said the girl with the package on her lap. "Cambridge," she repeated with a slight slur, "is a very good place to get the fuck out of."

"Oh, don't say that," said one of the sailors, slipping an arm behind her waist. "I love this town. The broads're beautiful. The bars're busy. Parley-voo français?" he said, pinching the girl's cheek. She laughed.

"Hey, what state is this anyway?" said the third sailor.

"The UnCommonwealth of Mass confusion," said the girl Dave had carried.

"Well, you've got a great town here," said the sailor with the cut lip. "Listen," he said, "we're not gonna hafta fill out a whole mess of forms and all that crap, are we? I mean it's not like we were at fault or something."

"I don't care what you do," Dave said.

"You mean you guys aren't the fuzz?"

"Do we look like the fuzz?" Dave asked him.

"Hell no," said the sailor. "You look like regular guys."

At Mt. Auburn they took the blonde down for X-rays. The other two girls went with her. The sailors were sitting on a bench out in the corridor joking around when the cops came in and headed over to the main desk. The sailors sprang up and began strolling nonchalantly down the corridor. Then like a flash they slipped out the door.

The girl Dave had carried came out. "Have you seen our friends?" she asked Dave.

"They took off," Dave said.

"Well, thank God; our husbands would have killed us."

It was a little after ten o'clock when we got back to the house. The others were all upstairs in the TV room. It was a gloomy bunch; the Rangers were leading by two goals at the start of the third period.

"I don't suppose you guys left us any food," Dave said.

"Need you ask?" said Kilroy.

We needn't have asked. In the kitchen there wasn't a trace of the cold cuts, and all that remained of Tom's pies were a few bits of crust. The refrigerator was empty. We had resigned ourselves to coffee when Jack Dillon appeared and motioned for us to follow him. He opened the door to the porch and pulled in a bagful of food.

"It nearly cost me my life," he said.

We put together huge sandwiches and made our way into the crowded TV room in time to see the Rangers score two quick goals back to back.

"Things were looking better," said Kilroy, "till you guys showed up." We were too busy gobbling down our sandwiches to bother kidding him back.

A minute later the bells were ringing: Box 735 in North Cambridge. Everything was going.

As we were racing through Porter Square the radio squawked: *"Engine Four to Fire Alarm."*

"Engine Four."

"Engine Four by the box. Nothing showing."

One minute later the radio squawked again: *"Engine*

Four to Fire Alarm. Apparently false. Give us the All Out."

"OK, Lieutenant. All Out on the box. Apparently false. Ten-thirty-six P.M."

Joe made a quick U-turn on Mass. Ave. The radio crackled again.

"Fire Alarm to Rescue Company."

"Rescue Company answering," said Cooper, picking up the mike.

"Respond to Harvard Square opposite the kiosk. Report of a bus accident."

The Rescue began the run down to the Square and we braced ourselves against the jolting bumps. We couldn't have been doing more than twenty-five, but on the icy roads it felt like a hundred. As the city flew past I had the strange feeling that we weren't a part of it. The city was made up of warm houses strung with colored lights, the tall Christmas trees in the Common, the wreaths hung across Mass. Ave., the brilliantly lit signs that read "Peace On Earth." With our flashing lights and screaming siren we seemed strangers in this place.

The thought of the bus accident made me suddenly uncomfortable. My stomach was clenched into a knot. I remembered how, when I first came on the job, I was afraid more than anything, more than of making a mistake, that I wouldn't be able to take it, that I would show my weakness in front of the others. It must be the cold cuts, I thought now; Jack Dillon's cold cuts and the motion of the truck.

As we rolled into Harvard Square we saw the police lights by the bus stop, but there wasn't a bus in sight. I was thinking it had all been a mistake; then I saw the crowd massed on the sidewalk in front of the Yard.

The cops cleared a path for us. "What kept you?" asked one of them.

"False alarm," I said.

"We thought you'd never get here," he said as we shoved our way through the crowd and burst into the center of the circle.

There were three young men stretched out on the sidewalk. "Oh, Jesus," I heard Dave mutter beside me. It looked like a massacre. One kid was cut practically in half; he lay writhing in agony with his guts hanging out on all sides. The second clutched his right arm; his hand was all but severed at the wrist. A river of blood flowed from him to the gutter. The third kid lay unconscious.

The crowd was excited. People were hollering and jostling each other for a closer look. As Dave and I knelt over the first boy they pressed in so close we hardly had room to move. The cops shouted at them and pushed them back.

The kid squirmed desperately, his blood smearing the pavement. He was a young kid, not more than twenty. I pinned him down and he shrieked. Part of his guts were touching the sidewalk. Dave scooped them up with gauze and moved them back over the middle of him. He cried out and his eyes rolled back and sank into the top of his head as Dave laid gauze over the wound.

Billy Stone came running with the orthopedic stretcher. We picked the kid up and carried him to the Rescue. Dave stayed with him. Billy Stone took the chair over to where Cooper was kneeling with the second boy, applying direct pressure to his bandaged hand. I took another stretcher to where the third kid lay. Joe was standing over him.

"I think he musta just passed out," said Joe.

We pulled out for Cambridge Hospital. It was crowded in the back of the Rescue, and dead quiet. The first kid was unconscious, his breathing regular but shallow. "Gimme the oxygen," said Dave. He placed the mask over the boy's face while I felt his neck; the pulse was still strong. One arm dangled over the side of the stretcher and when I placed it against his side I saw that blood was oozing out from under the bandage.

"I don't get it," said the second kid. His face was pale and taut, the bandages on his wrist soaked dark with blood. "I don't get it at all. He never gave us a chance. We didn't do nothin'. We were waitin' for the bus and he just come and says he wants our money. Rudy says we don't have none. Then he pulls out this knife, this fuckin' banana knife, and puts it right into Rudy. Carl takes one look and goes out like a light. Then the guy turns to me and says he wants my wallet. I'm holdin' my hands up in front of me as I reach for it, but he didn't wait. I never even seen it comin'. He never gave us a chance."

We took them into the Emergency Room and went to wash up. A few minutes later the nurses wheeled two of them quickly up the corridor.

"What's the story?" Cooper asked Mary Lane in passing.

"It's a mess," she said. "The one with the gut wound is critical. The other one'll be lucky if they save his hand. They're both going up to surgery right away."

"How's the third kid?"

"Shaky. But there's not a mark on him."

Cooper filled out his report at the desk and we picked up our gear and took it out to the Rescue as a squad car pulled up. Cooper knocked on the window. The cop rolled down the window a crack.

"Say, Barney," Cooper said, "what the hell happened down there anyway?"

"I don't know," he said, "but it just come over the radio they got the guy that did it. Frank Granatino caught 'im. That Granatino's the luckiest sonofabitch I ever seen. He's off duty and he's standin' in front of the Wursthaus there and he sees this guy runnin' about a block away. So he decides he's gonna chase 'im. Christ almighty, Billy, if we stopped every fruit that was out jogging at night in this town we'd be haulin' in half of Harvard College. But it turns out to be the guy. He's twenty-two years old, they think he's on drugs or somethin'. Howdja like it?"

"I coulda done without it," said Cooper.

"Yeah," said Barney. "But it sure was a lucky break for that Granatino."

I don't understand cops. We see a lot of them in our work, and of course we know a lot of them from school. Some of the things they do and say don't make me too happy. But then they've got a very different kind of job and different motivations. Firemen are motivated by the lifesaving impulse, while cops are dealing continually with a criminal element they can't trust. Maybe if I was doing that day in and day out I'd get to feel the way they do about a lot of things; but I don't think I'd ever want their job. I suppose I understand as little about what makes them tick as most people understand about us.

We headed back to the station. It was odd to see so much traffic at that hour. It was almost midnight and church bells were tolling. If I hadn't been working Bev and I would have been on our way to Mass.

My mother never missed a midnight Mass on Christmas Eve. I remember checking the closets with my older brother, Ray, as soon as she left, to make sure our stuff was still where she had hidden it. We used to leave coffee and cookies out for Santa Claus and we could never get to sleep. The lights on the street were beautiful, like jewels, and once I remember looking out in the middle of the night and seeing enormous flakes of snow quietly falling.

Nine o'clock Mass was the children's Mass at St. Paul's. I was an altar boy, but it wasn't too bad, because almost everybody else in my sixth grade class was one, too. As an altar boy you'd get out of school a couple of hours a week, and at a funeral or wedding Mass you could pick up as much as fifty cents, but that only lasted until the nuns found out we were spending the change at Cahaly's Delicatessen instead of kicking it back into the poor box. When we weren't on duty at the altar we used to cut a lot of Sunday Masses; the trick to that was finding out what color vestments the priest was wearing so you could answer the inevitable question on Monday morning. We were expected to be in the choir, too, which meant daily practice after school. The only way to get out of that was to pretend your voice was changing, but fooling the nun took a lot of practice itself. She'd put her ear right up against your mouth and if you failed to convince her she'd make you very sorry.

We tried to get out of everything, but there was no getting out of nine o'clock Mass on Christmas morning. My mother would come and sit in the back and watch me be an altar boy. Then she would wait for the eleven o'clock Mass, when I sang in the choir.

My father was a traveling salesman and he spent a lot of time away, but he'd be home for Christmas dinner. First we had lasagna, then turkey, then salad, then we'd have seconds. My father would unbutton his jacket and when he had had his fill he'd take Ray out for a walk. My grandfather would break out the fruits. He grew up somewhere in Russia or Poland, I can't remember which, and he told me the winters there were so cold that if you threw a pan of water out the door it froze before it hit the ground. At night they heated rocks and wrapped them in sacks and took them to bed. One of the best things about Christmas, he said, was having an orange. His family sat around the fire and peeled the orange carefully and each person got one section of it. At Christmas, he said, you believed that things would get better.

٥   ٥   ٥

The firehouse was dead quiet. Dave had flecks of blood in his blond hair and his pants needed changing. Cooper went to write up his report and turn in. Joe and I had a cigarette up in the kitchen.

"Wanna watch the tube?" he said.

"No."

"How about some Ping-Pong?"

"No thanks."

"I'll spot you ten points."

"Forget it, Joe. I don't want to have to beat you."

Joe poured himself a cup of coffee. It was old and too strong and he poured it down the sink. "Jesus, I hate this night," he said. "I'd rather work New Year's Eve, you know, and be home tonight. What the hell is New Year's anyway? You just get drunk and pay for it the next day."

Billy Stone came in and got some milk out of the refrigerator.

"How about a quick game of Ping-Pong?" Joe said.

"Nah, I'm going to sleep," said Billy. "I gotta drive out to Springfield in the morning."

"I'll give you five points."

"All right," said Billy.

I went down to our dormitory on the second floor. Dave was lying in bed, staring up at the ceiling, his hands folded under his head. I sat on the edge of my bed and pulled off my shoes.

"What's the matter with you?" said Dave.

"Nothing's the matter with me," I said. "I'm going to sleep."

"Well, you look terrible," he said.

"I don't think Dillon's cold cuts did me any good."

"Or that kid's guts."

"They didn't help."

Dave threw off his blankets and sat up in bed. "Fuck

this, Lar," he said. "I'd really like to get me the Christ off this fucking zoo."

Just then Billy Stone and Joe stormed into the room.

"What happened to the big match?" I said.

"Ah, the stupid ball got busted," said Joe.

"How'd that happen?" I asked him.

"The stupid ball made the mistake of rolling under his foot," said Billy.

"Get me outta here," Dave groaned.

"What's that, Swanson?" Joe said.

Dave pulled the pillow over his face and shouted a few obscenities into it.

"Poor Swanson," Joe said. "It must be that time of the month again."

Dave threw off the pillow. "Look," he said. "I've had five years of this. I'm putting in my slip in the morning."

"And where do you think you're going?" said Joe.

"Up on the hill by Huron Ave. I know a nice little Engine Company up there where you don't have any of this junk and you spend your spare time planting flowers and fixing kids' bicycles."

"You'd hate it," said Joe.

"I'd be happy as a pig," said Dave. "It's what I'd really like."

"What I'd really like is a Ping-Pong ball," Joe said.

"Yeah? Well, maybe Santa will bring you one for Christmas."

"Nah," said Joe, unbuttoning his shirt. "I ain't been good enough."

"Well, maybe," said Dave, "if you hold your breath and lie very still and wish very hard, you just might lay one."

"And you know what you can do," said Joe, unbuckling his trousers.

"Hey," said Billy Stone, climbing into bed. "You know what I'd like? A little peace and quiet."

"Amen," I said.

I draped my shirt and pants across the back of the chair beside the bed, set my shoes and socks on the seat and slid into the cold bed. Joe shut off the light. The brass poles gleamed in the light from the hallway and the wind rattled the windows. The clock on the wall read one-fifteen.

I thought about Dave. I wondered what had kept him on the Rescue for over five years. He was always saying he'd had enough, but he had never turned in his papers, and on the surface he seemed the one least affected by it all. I thought I understood the others better. Cooper and Stone were more like me; they got a deep personal satisfaction out of the Rescue. Cooper probably worried more about doing a good job, but then he was Lieutenant and had the responsibility for the Rescue Company when he was on duty. On the whole I think he coped with it the best. Joe was more of a fire fighter than the rest of us; he liked being on the move, he saw the risks as a challenge and he took great pride in doing a technically competent job.

But Davey I didn't understand. Usually he just let things roll right off his back, and you could never get him to talk about it, really. I wondered if he had been any different before getting on the Rescue.

I thought of the stupid argument I'd had with Bev over the vitamins and the way I was getting to be overprotective of Larry. I didn't see how I could help it. People know in the backs of their minds the bad things that can happen, but I really *see* them. I've *seen* what happened to the kid that chased his ball into the street. I've seen what something as commonplace as a car door does to a kid's hand. The things that could have been prevented make me mad, and leave me irritated with people who may not be at fault, and who at any rate can't be perfect. All of it begins to distort your personality—you become preoccupied. It's hard to remember how I was before, but I think I used to laugh and joke more.

Even that doesn't matter. What matters is my family. As long as I have them to love and enjoy I'm all right. But to enjoy them freely you need an illusion: you need to believe that nothing bad will happen to them. And the trouble is, what happens to other people happens in part to me; what hurts other people destroys the illusion that I need in order to live normally.

Enough. Davey's right to keep it to himself the way he does. And it's not good to try to think in the house at night. Your thoughts rarely make sense in the morning.

In the distance I heard the long, slow ring of the night phone. It sounded very far away. I got up. When the phone rings at night it's almost always for us. Cooper was yelling from the officers' dormitory down the hall. The night lights flashed on as the others rolled out of bed, shielding their eyes momentarily. We pulled on our clothes and hit the poles.

The streets were deserted. It was just a little before two-thirty in the morning. Fire Alarm had said the call was for a sick woman. We swept downhill toward East Cambridge, running a long string of red lights. We raced past the rickety, flat, three-decker houses and the vacant, littered lots of industrial Cambridge, bouncing hard over the potholes and the railroad tracks. Going from Harvard Square to the eastern end of this town is like moving into a war zone.

It took us six minutes to reach our destination: a lone brick tenement standing between a truck depot and the railroad yards. Lights showed in the front windows. As we climbed down from the Rescue a Puerto Rican in a huge sweater came running toward us, waving his arms wildly and shouting in Spanish.

We followed him into the hallway, carrying the chair

and the emergency box. Neighbors crammed the narrow
stairway. We climbed to a third-floor apartment. In the front
room two small children in pajamas stood on a bare mattress.
There was no other furniture. In the back room a woman,
about thirty, lay unconscious on the bed. There was a cross
on the wall above her. She had long black hair, olive skin and
a perfect round mouth. She was very beautiful; but the cast
of her face told us she was dead.

The man held out his hands to us. "Please," he cried.
"Please."

Cooper lifted her eyelids: the pupils were fully dilated.
She had no pulse, no respiration. Her ears were bluish. We
lifted her onto the floor and Dave knelt over her and started
cardiac massage; Joe ran down for the stretcher and the
board.

"Give me an airway," said Cooper. Billy Stone handed
him the plastic tube from the box. Cooper tilted back her
head and inserted it into her mouth. He held up her chin
with one hand and cupped the other tightly over the facepiece while Billy Stone began ventilation with the ambu bag.

"How long has she been like this?" Cooper asked.

The man shook his head.

"Lar, see if you can find a translator," said Cooper.

"I speak," the man said. "Her heart not good," he cried,
clasping his hand to his chest. "The doctor, he was jus' here."

"How long ago?" Cooper said.

"Today. This noon."

It didn't look good. Her heart and breathing had
stopped—she was clinically dead. The cells of the brain can
live only four to six minutes from the time the blood stops
circulating; then biological death starts to occur. After that,
even if you bring them back, they're just vegetables. To
stand a chance of success, a resuscitation effort must be

started within this four-to-six-minute period. But there's no way of telling when this period has begun.

Her husband stood sobbing against the wall, his arms wrapped around himself, his body shaking convulsively. His sweater was at least three sizes too big and covered his hands. The two children gazed at their mother from the doorway, their eyes black as coal. "Get those kids outta here," Cooper said. I led them into the front room, their tiny hands lost in mine. Two cops were coming up the stairs and strange faces crowded the doorway; one of the women stepped forward and took hold of the children.

I went back in and the man grabbed me. "Angelina leev," he shouted, shaking my arm. "No die." His eyes, red and brimming with tears, were riveted on mine. He was a wiry, dark-complexioned guy. His nostrils flared in and out like those of an excited horse; he was wound up tight and ready to explode. He looked at me, daring me to say it, but I didn't know how to tell him. He shook my arm again. "No problem," he said anxiously. "No die."

I turned away from him. He came after me. One of the cops pulled him aside. He started to scream. "*Please,*" he shouted at me. "*Please!*"

I dropped beside Dave and took over the massage as Joe brought in the stretcher and the board. "It's no use, Larry," Dave said, but he let me work. Billy Stone was on the bag, Cooper securing the facepiece. It had been at least two or three minutes already. The husband was hysterical and the children were crying in the front room. We had made enough of a show of it. Dave and Joe slipped the board under her and began to lift her onto the stretcher, while I kept working.

As I fell into the rhythm of the massage, pressing down sharply with the heel of my hand once each second, the old feeling of disbelief came over me. You've seen so many dead people and you know death when you see it, but still you can't accept it. I had the feeling that if we could only work hard enough, *want* hard enough, we could bring her back. Her robe had fallen open; her breasts were small, like those of a young girl. She was still warm. I heard her husband sobbing. Her face was so composed. My cheeks burned. Like him, I couldn't believe she was dead.

Cooper bent over her and examined her eyes again. I looked carefully at her: her earlobes were pinker; her color *had* come up. "The pupils are coming down," Cooper said. "Let's go!"

Dave tapped on my shoulder and relieved me on the massage; I helped with the stretcher. One of the cops led the husband down and the others cleared a path into the hallway. We tilted the stretcher to navigate the stairs. One of the neighbors held onto the two children. Glancing back into the apartment I saw a small Christmas tree in the corner, and underneath it a few presents and one box that lay open.

The cops took the man to one of the squad cars. When he saw us emerge from the building and head for the Rescue he broke away and ran for his wife. Two of the cops cut him off and grabbed him by the arms. He lunged for his wife, struggling like a crazed animal. He wanted to come. I tried to tell him that the police were taking him to the hospital, that he couldn't come with us in the truck. He demanded to come. We didn't want him in the Rescue. He could get in the way and it wasn't a nice sight. But Cooper must have seen the look in his eyes. "All right," he told me, "let him come."

The man climbed up into the Rescue and we pulled out, Dave keeping up the massage, Billy the ventilation. Cooper made his way forward to radio a Code 99. The guy sat still, realizing now that we believed there was a chance. He bowed his head and started mumbling prayers. His dark eyes burned and tears streaked down his face. Cooper relieved

Billy Stone on the ambu bag and I relieved Dave on the massage.

The doctors and nurses were waiting when we rushed her into the trauma room. I kept up the cardiac massage as they slid her off the stretcher onto the table. They slid the endotracheal tube down her throat and fastened the EKG bands around her wrists and ankles. One doctor plunged a giant needleful of sodium bicarbonate into her stomach; another took an EKG reading. "Ventricular fibrillation," he shouted. Somebody rolled the defibrillator over to the bed. Her heart muscle was quivering wildly; they had to try to shock it back to a normal beat. They greased her chest and the defibrillator plates in preparation for the shock. Somebody took over the massage for me and I left the room.

Twenty minutes later they were still at work, but there were no signs of life. It was pointless now, and they wouldn't keep it up much longer.

Avoiding the waiting room, where I could see her husband pacing up and down, I stepped outside for a breath of air. It was three-thirty in the morning. The air was perfectly still and the whole world seemed frozen fast. The stars looked enormous. I pulled up my collar and walked around the half-empty parking lot, the thin crust of snow crunching underfoot. In a few places there were smooth sheets of ice and if you stomped on them once they would shatter around you.

I heard rapid footsteps; it was the husband. "Hey, you," he called, grinning. "*Gracias*. I like to thank you very much."

He stood there beaming at me, the sleeves of his dazzling sky-blue ski sweater hanging down over his wrists. His eyes were shining and a gold tooth gleamed in the corner of his mouth. He was a good-looking guy, and now that he was smiling I saw that he really wasn't much older than me.

"Look," I said, starting to tell him; but I couldn't find the words.

"Angelina," he said. "I tell you, Angelina, she not die. You don't believe me, OK. But now," he said, tapping his forehead wisely, "now you see. A young woman like that," he said, flourishing his hands in an eloquent appeal, "how is it possible she can die?"

I said it was too cold to stand still and started walking. He was at my side. "I'm not cold," he said, patting his sweater. "She has given me this tonight," he told me. "It's a little big, no? But jus' you wait, she will fix it nice."

Light spilled out of the front entrance as the door banged open. A nurse poked her head out. "Mr. Mendez?" she called.

"I go," he said, grabbing my hand and pumping it once. "She need me now." He turned and broke into a trot and disappeared into the building.

FOUR

IN THE LONG HALLWAY OF THE HOUSE WHERE I grew up, the doctor unlocks the door and asks me if I would mind trying on something for him. It's a leg made of rubber. It rolls up like a boot and fits snugly around my thigh.

In the bedroom I ask my father how I lost my leg. He walks out and a voice says "Yes." A man with thick glasses and a huge nose approaches, holds a mirror up to my face and breaks into hysterical laughter. I start to slug him. Black liquid runs from his eyes. He falls and I drag him between the beds. I try to push the body underneath one bed but there is already something under it.

A cop escorts me to the truck and starts to fasten me to the chair. I tell him I can do it myself. The door bangs shut and the echo reverberates through the truck. As I strap myself in I notice a sheet-covered body on the stretcher beside me. My father starts to pull the sheet down. I ask him to stop. The cop turns his back. I tear at the straps, shouting at my father, but he won't stop. I break free and start to run. I'm running but there is something chasing me, right behind me. My leg grows stiff and wooden and I can't run any more. I turn and as I start to slug, I scream.

"Larry, are you all right?" Bev was asking me. I was in a cold sweat and I realized that one of my legs was asleep. I had a headache. "Was it that dream again?"

"Sort of."

"Did you see what it was?"

"No, I never saw it; I just turned around and let loose."

"I'll say you did," she said, sitting up. "You nearly bounced me out of bed. Did you get some sleep at least?"

"Sure."

"Sure you did," she said. "You couldn't touch your supper but you had to have a pizza before bed."

"I wasn't hungry at supper."

"And it's no wonder, the way you smoke and drink coffee."

"It's been cold. I like a lot of coffee when it's cold."

She got out of bed and turned on the lamp. Under the blanket I massaged my leg. It was alive with pins and needles.

Bev put on her robe. "Larry," she said, "you promised me you'd make that appointment."

"I've had a cold, that's all. What's he gonna tell me?"

"I don't care," she said with a tense edge in her voice. "I've had enough of this. Have you taken a good look at yourself lately? You look lousy."

"All right," I said. "I'll do it if it'll make you happy."

"It will," she said.

Two days later, on a freezing, slushy-gray February afternoon, I went to see the doctor before work. He was a guy Bev's sister had recommended, and he had a ritzy office down in Harvard Square. There were soft leather chairs that squished when you made the slightest movement, romantic music piped in through loudspeakers hidden in the tropical palm plants, and a harem of young assistants slipping silently in and out of the doors.

The blonde behind the desk was beautiful. She had me fill out a card.

"Funny," she said, looking at the card, "you don't look like a fireman."

"What do they look like?" I said.

"Oh, you know," she said.

"Yeah, I know," I said. "They wear helmets and red suspenders, and they're always looking for something to mash with their axes."

She laughed. Listen, she told me, it was no joke. She had a girl friend who had had a small fire in her studio apartment over in Boston, on Beacon Hill. It was just a tiny little fire in the kitchenette, but they broke all her windows and put holes in the roof, walls and ceilings.

I gave her a quick course in fire-fighting strategy. I told her that the main obstacles to rescue work and fire fighting are smoke and heat, and how ventilation helps remove these obstacles. Opening windows, or knocking them out when they're stuck, reduces heat and smoke. Chopping a hole in the roof at precisely the right spot draws the fire upward, minimizing the risk of back draft, an explosion produced by the heat of the fire. It also decreases the amount of mush-rooming, a condition in which a fire deflected from its nat-ural path upward moves sideways and down until it spreads throughout the entire structure. Ventilation is by far the lesser of two evils. At the cost of a few broken windows and a damaged roof, it makes it possible to save a building that might otherwise be lost, and to reduce the amount of water damage that would result from a more protracted battle. As for tearing out walls and pulling ceilings, I explained to her how a fire gets behind surfaces and extends upward, and how you have to make absolutely sure you've extinguished all of it.

But I must not have done too good a job of explaining, because there was no convincing her that firemen aren't a bunch of misfits who get a secret thrill out of tearing things apart. In her defense, she had plenty of company; people just don't understand what fire fighting involves.

"They wrecked her whole place for nothing," the girl concluded, reaching back to smooth down her hair. "I mean, it's nothing against you personally. I'm sure you don't do that over here in Cambridge."

"Well, that's certainly true enough," I said. "We don't want that type of wild man here."

"I'm glad to hear it," said the girl.

"Our boys just like to play a little with the trucks and ding the bells," I told her. "You know. Ding ding ding." I flashed her a grin and sat down. A few minutes later when one of the nurses called me in, the blonde gave me a funny look.

The doctor examined me. He was an elderly gentleman with a heavy European accent. He couldn't find anything wrong mit me except a slight veezink in the chest. I told him I'd had asthma as a kid. Really, he said, that's very interesting. Had I always had it? Yeah, I'd always had it. Ya, well, he said, I was probably just feeling run down. It was probably a touch of the virus. Everybody had it. It was the extreme colt mit the dampness, he said. He wasn't feeling so hot himself. He gave me a prescription. Just take it easy and keep warm and dry, he said, patting my arm as he showed me to the door. He was a real con artist, this guy. Come back in a few weeks when you're still feeling bat, he said.

The blonde behind the desk smiled warmly when I came out. She wanted to know whether I would like to pay then or have them bill me. I told her she could bill me. She didn't look that great from close up after all.

I slogged through the ankle-deep slush in the Square and cut through Harvard Yard to the firehouse. The paths were well plowed in the Yard, and Widener Library was lit up like an ocean liner at sea. Night was closing in and the wind rattled the bare trees overhead.

"Stay warm and keep dry." It had been a hell of a month for that. I'd been wet and cold since the fire the week before. Everybody had said that was the coldest night they'd ever spent on the Department; even old Jack Dillon couldn't remember worse. Just thinking about it still gave me the chills.

The fire had broken out around eleven o'clock in a big old abandoned house that belonged to Harvard, just around the corner from the station. When we pulled out of quarters onto Cambridge Street we could see it clearly. Flames were pouring out of the basement and the first-floor windows. We were the first company off, and we made it into the front hallway. The cellar was fully involved—we couldn't get down the stairs—and the first-floor rooms were going all around us. We climbed to the next floor to search for victims and check the extent of the fire. But the fire came up the stairs after us, forcing us out onto an icy balcony.

The second alarm had been sounded. Apparatus crammed the street below, their red lights flashing. Engine Companies were laying line, truckmen carrying ladders to the building. From where we stood they looked like soldiers mounting an attack. It was freezing cold and there was a powerful, shifting wind. Smoke was tumbling out through the windows, and flames swept across the room behind us. Joe, Billy and I yelled for a line. Davey, who'd been driving, came up over a ladder.

"What's holding up the lines?" Joe said.

"The nearest hydrant's frozen," Dave answered. "Come on," he said, reaching for the door. "What are you waiting for? Let's get in."

"Be careful," I told him. "The place is full of fire."

Davey looked at me like I was crazy. He pulled open the door. There wasn't a wisp of smoke: the wind had shifted.

We followed him back into the house. The walls and ceilings were scorched. We could see the glow of fire out in the hallway. There was a distant rumbling and trembling from below, like a ship's engines starting up. Tongues of fire shot through the doorway and we were enveloped in smoke. We were forced back out onto the ledge as the flames roared across the room.

If we had had a line we could have beaten the fire back. You can knock down a lot of fire in an abandoned building. The fire doesn't build up to as great an intensity, because there isn't any furniture to feed on, and the ventilation problem is lessened because the windows are usually broken. Also you can use a two-and-a-half, because you don't have to worry as much about water damage.

But without a line we couldn't do anything. We retreated down the ladder. As we hit the ground somebody shouted for the Rescue. They were carrying a man—the Lieutenant from Engine Nine—out the front door. Part of the ceiling had let go down in the cellar.

They looked worried as they laid him o1 the ground. His metal helmet was creased but there wasn't a mark on him. He was just stunned; he started coming out of it immediately.

The men were excited and angry. Somebody said it had to have been a torch job to have gotten rolling so good. Somebody else said that it was a lot easier to burn down a building than to wreck it.

We stretchered the Lieutenant and ran him down to Cambridge Hospital. They sent him up for X-rays right away. We didn't have time to hang around for the results.

When we returned they were setting up the ladder pipes in the street and settling in for a long siege. The fire had broken through the roof and the building was ablaze. Flames were showing in all the first- and second-story windows and shooting up above the roof. It was impossible to get inside with the lines for very long. We would work ourselves in but the wind would shift and force us back out, and you can't fight a fire from the outside.

It was cold out on the street. The temperature was five above zero but with the twenty-mile-an-hour wind the chill factor made it feel more like thirty below. Gusts of wind howled up the street, driving the spray back on us. The water froze almost instantly. Icicles hung so thick from the rims of our helmets that you couldn't tell who was who. Our rubber coats were quickly glazed with ice. The wind cut through our wet gloves and stung our fingers. My ears ached and when I closed my eyes tight I saw an enormous purple disc. It was cold beyond hand-slapping and foot-stomping. It was cold beyond shivering. I was a walking block of ice and after a while I didn't care. All I wanted was to curl up someplace and go to sleep.

The Deputy kept spelling us and, since it was just around the corner, sending us back to the house for dry clothes. We'd yank ourselves out of our coats and they would stand up by themselves against the wall. We'd lean over the radiator and get as much steaming coffee into ourselves as we could. Our cheeks and fingers, which were numb when we came in, would begin to tingle as we thawed out, and our teeth would start to chatter. Then it would be time to go back out. The guys we relieved would be so cold and tired we'd have to tell them twice before they'd give up the line. In five minutes we'd be wet again. In ten we were frozen solid.

At one point I was jockeying line for Engine One with Billy and Dave when Joe trudged over, his hand under his coat like Napoleon. Tiny icicles clung to his mustache, and his lips were blue.

"What the hell are you doing?" said Dave.

"Warming my fingers," Joe said. "I think I got frostbite."

"Frostbite?" said Dave.

"Yeah," said Joe. "My glove got wet." He pulled his hand out from under his coat. "They feel sorta doughy," he said. "When I touch them I don't feel nothin' and they don't get red."

We all looked at his fingers. I couldn't see anything; they looked normal to me. In front of us the house was blazing away and the street, glazed solid with ice, shone like a sheet of copper in the glare of the fire.

"They're just cold," said Dave. "Rub 'em."

"Rub 'em?" said Joe, dumbfounded. "That's brilliant, Swanson. Didn't you ever see *The Alaskan Mummy?*"

"No," said Dave. "I don't watch that kind of stuff."

"Well, as a matter of fact, you do. You were sittin' right beside me when we seen it. Don't you remember the part where the German scientist kills the Russian general by having the Eskimo massage his frostbitten limbs?"

"No," Dave said, snapping off an icicle that hung from the side of his helmet.

"Well, you don't rub frostbitten tissue. It contains sharp ice crystals that cut. Until you can get medical attention you're supposed to warm frostbitten fingers by blowin' on them or holding them motionless under your armpits."

"Yeah," said Dave. "Well, when your armpits cool down I know another place you can stick 'em."

"Look," said Joe. "I don't got time to shoot the breeze with you. Cooper wants one of you to drive me down to the hospital."

"I'll go," Billy Stone volunteered.

"No, I'm driver tonight," said Dave. He threw a big arm around Finnegan's shoulder. "Joe, sweetheart," he said, "why didn't you just tell me you wanted to go for a ride?"

The Rescue pulled out, marooning Billy and me. We

wandered from company to company, taking our turns on the lines and the deck guns mounted on the hose wagons. It was hopeless. The wind kept feeding the fire. No sooner did we get some of it down than the wind blew new life into it and swept it across. There was no putting the damn thing out. For four hours we poured water on it. For four hours the wind threw the water back. The big ladder truck, which had been parked in an alley downwind of the building, was caked solid with ice; when we were finally ready to leave the scene it had to be towed back to the house.

The next day they had a picture of the fire in the paper. They called it "Winter Wonderland." The shrubs were encrusted with layers of ice and looked like miniature frozen waterfalls. The branches and twigs of the trees were icy and delicate as crystal. There were icicles strung as carefully as tinsel from the wires overhead. It was a beautiful sight when you saw it in the Thursday papers.

Then, over the weekend, we had a blizzard; nearly two feet of snow in all. The snowflakes were still swirling down when our group arrived for work Sunday morning.

The Engine and Ladder Companies went to shovel out the fire hydrants, while we shoveled the walks and the apron in front of the house. We had fifteen calls: automobile collisions, mostly, and elderly ladies who hadn't been to Mass in twenty years suddenly having to go and falling on the slippery streets. And it was tough getting around. Cambridge, with its old narrow streets and high density of population, is the third most congested city in the United States. Traffic was at a virtual standstill, with cars double-parked, cars stalling out and cars creeping along buried under a foot of snow with tunnels dug to the windows so the drivers could see out. It was cold and wet and I hated the sound of the chains going chicka-chicka-chicka as we churned through the slushy streets.

Even when we did get back to the house it was impossible to escape the cold. The huge, drafty old building was hard enough to heat under normal conditions, but Ladder One was still frozen solid from the fire, a giant block of ice that turned the apparatus floor into a freezer and lowered the temperature throughout the house. Even with the radiators sizzling away the thing hadn't thawed out. It looked every bit as frozen as the night we'd towed it back.

On Monday things let up a little. But just before our tour of duty ended we were called down to Memorial Drive where the Anderson Bridge arches over the frozen Charles. A sports car running a red light on the icy pavement had collided with a huge tanker. It hadn't been much of a contest. The truck was sitting on top of the car when we got there. The car looked as though it had gone through a compacter. The top was flattened and the doors had buckled in and folded up over the roof. The cops guessed the kid must have been doing at least sixty when he hit the intersection.

We attached the come-along—a heavy chain-and-pulley rig—to the underbelly of the truck. It was hard to get leverage and we were under there for a good half hour. The wind whipping off the river numbed our faces. Finally we pulled one of the doors free and slid the guy out. There wasn't a mark on him. He'd been killed on impact.

It had been some week for taking it easy and staying warm and dry.

Arriving at the house for the start of the Wednesday night shift, I felt exhausted. I hoped for a slow night. Things had been happening too fast; it seemed to me that we'd been on the run for months. It had been a hell of a winter.

The evening started quietly: a woman who had slipped on her porch steps, a man who felt pains in his chest, and two false alarms. At eleven o'clock we received a police call to go down to the Central Square subway station. There was an injured person on the outbound side.

As the Rescue sped down Broadway Joe called up to Cooper, "Did they say what the injury was?"

Cooper shook his head.

Billy Stone and I filled our pockets with gauze and bandages. It's bad when you don't know what you're going to: you're forced to prepare yourself for the worst.

"I hope this isn't gonna be a jackin'," Joe said.

Billy was riding standing up, his arms braced against the cabinets. He looked like a parachutist about to jump.

"Would ya mind sittin' down?" said Joe, his foot tapping nervously. "You're blockin' my view."

"Why should I?" said Billy. "There's nothing to see anyway." But he sat down.

We rolled into Central Square and pulled up behind the police cars. Passersby lined the sidewalk beside the subway entrance. Billy started pulling out the big canvas bag with the twenty-five-ton jack and the wooden block, which was stored in the outer compartment. Dave leaped down from the cab to help him, while Joe got out the third-rail tester.

Cooper and I ran clattering down the two flights of metal stairs. The damp, metallic smell of the subway rose up to meet us. Through the iron railing we could see the darkened train sitting in the station, its doors open and the people inside. There was an eerie, tomblike silence.

The MBTA inspector let us through the turnstile. "He's under the front car," he said.

"Is the power off?" Cooper asked him.

"All set," he said. "What's your name?"

Cooper told him his name.

The official would call up the power-control office and give them Cooper's name. They wouldn't put the power back on until Cooper ordered it personally.

We ran down the platform past the cops, toward the rear of the three-car train. There were a dozen or so people on the platform and one of them, a tall woman in a red coat, shouted that the man was at the other end. Cooper waved her off. There was no way that we could squeeze under the front-wheel housing. We had to crawl the length of the train.

Cooper and I jumped down onto the tracks. Joe was right behind us with the third-rail tester, a light bulb enclosed in a steel-mesh cylinder. The inspector had said that the power was off, but you couldn't play it safe enough. A train coming up from behind could set up an electrical override to the dead section of track. There was supposed to be some device to prevent this, but who knew what was happening a half mile down the tunnel.

I stared at the third rail as Joe attached the wires to it. The rail was black, unlike the shining silver of the tracks polished by the passing trains. Staring at the rail, I remembered suddenly the face of a girl from another time. She was twenty-one and one night she wrote a note saying she couldn't live without some guy. Then she came down here and kissed that rail.

Cooper switched on his hand light, lay down in the shallow pit between the two tracks and slid under the end car. I crawled in after him. Joe stayed behind to test the rail.

The ground was hard and damp. I crawled with my stomach pressed against it. The ties embedded in the pit scraped my belly as I moved. Just above my head I could feel the bulk of the car. Between the underbelly of the car and the rails all I could see was the grimy base of the wall. The pit was filthy with grease and cigarette butts, and the dust we kicked up filled my nostrils with the stench of the subway. I'd hated that smell from the first time I'd walked past the subway grates and heard the rumbling of the trains below.

Now it was dead silent except for Cooper's shoes scrambling in front of me, the sound of my own breathing as I crawled and, far off, a sharp drip-drip-drip of water. No matter how many times we'd done it for real or in drills, I couldn't get used to the feeling of crawling beneath those cars. I hated the smell and I hated the idea of what we were going to find.

Between the third and second cars there were hoses that hung down under the coupling. I hesitated, pressing myself flat against the pit, before snaking forward. Drops of liquid fell onto my cheek and ran down into the corner of my mouth; it tasted metallic, like water from a canteen. Then suddenly—SSSHHH—the air was electrified with sound. My head flew up into the train and a blinding white light flashed before my eyes. My ears buzzed.

"You all right?" Cooper shouted.

"I don't know," I said, trying to find my voice. I realized it had only been the pressurized air of the door, but I couldn't stop my heart from hammering. "Sonofabitch," I said. "I nearly jumped out of my skin."

"That's nothing," Billy said. "I made it clear up to Mass. Ave. and back." Then we were crawling forward again.

Midway under the second car we paused for a second to catch our breath. Cooper held his light in front of him and played it up ahead. "There he is," said Billy. The man was under the first car. All you could see of him was his boxer shorts; he looked like a heap of abandoned clothing.

I put my head down on my hands. It was throbbing painfully and I could feel the blood pounding in my temples. He'll be mangled bad—that was all I could think.

"All set?" said Billy.

"Let's go," I said.

We slid forward, squeezing under the coupling between the first and second cars. We could see him clearly now. There was the sound of dripping again, louder this time. Then we heard a low moan. "Oh, Jesus, no," cried Cooper. "He's still alive."

We slithered forward fast. If he was under the wheels we'd have to jack the train off him. But that wouldn't take too long; we would only have to jack it a quarter of an inch to slide free what was left of him.

Billy got to him first. The man lay between the rails, naked except for his shorts. His body was intact. There was a deep gash running from the top of his head down between his eyes, but his pulse and pupils were good. Aside from his head wound there wasn't a scratch on him. He was a tiny man, about forty years old. He couldn't have weighed more than a hundred pounds. His size had saved him. It was a miracle the wheelhousing had cleared him at all.

"We'll take him out with a blanket," said Cooper. I yelled down to Joe at the far end of the train. Cooper swathed the guy's head completely in bandages. Davey came crawling with the blanket; he couldn't believe the guy was still alive. Cooper and I slid the blanket under him; then, crawling side by side, we dragged him down the length of the train.

Joe and Billy Stone were waiting with the orthopedic stretcher on the tracks. "You mean he's alive?" said Joe. We picked him up and lifted him to the platform. There were a lot of people crowding about and the cops kept them back.

"Where the hell are his pants?" said Joe.

"Who knows?" said Cooper, his face and uniform smeared with dirt and grease. "They weren't under the train."

"He wasn't wearing pants," said the woman in the red coat. There was an elderly man at her side; they both seemed upset. "He just came in like that and he got down on the tracks."

"He didn't jump?" said Cooper.

"No. He just lay down," she said. "We—this gentleman and I—we tried to reason with him, but he wouldn't get up. There wasn't any time. It was awful."

"He must be a crazy one," said the old man. "How is he?"

"He's alive," Cooper said. "He wouldn't have been if he'd jumped."

"It's a miracle," said the woman.

"Yeah," said Billy, "I guess it is."

We started up the stairs with the stretcher. One of the cops came running up after us and said they'd just found the burned remains of his clothing in the men's room. There was a card from the Veterans' Hospital in his wallet, and they were radioing for details.

On the way down to Cambridge Hospital the snowbound city seemed flooded with light. Yes, it's a miracle, I thought; but it didn't make me feel good. Two kinds of people attempt suicide: those who are confused and are crying out for attention, and those who really want to die. The first kind usually leave the door partly open, and there's some hope if you save them. But anybody who lies down in front of a train wants to die. And if by some fluke the door was left open today, it would be closed tomorrow.

At the hospital they told us he had been identified as a psychiatric patient at the Veterans' Hospital. He'd been reported missing.

In the Rescue, returning to the house, Joe started telling Billy about some guy they'd picked up a couple of years back on the railroad tracks near Somerville.

"Were you on then, Lar?" Joe asked.

"I don't remember," I said. I had a headache now and the top of my head still throbbed from the jolt it had taken. The front of my clothes was slick with slime and filth.

"You weren't," said Joe, "or you woulda remembered." He turned back to Billy. "Well, it was up near Porter Square there. The guy was in five pieces. The head and trunk was together, but the rest was scattered. We hadda go huntin' under the train for the limbs. We put 'em in a body bag. It was a real mess. George Benson was Lieutenant then.

Was George still Lieutenant, Lar, when you come on?"

"Yeah," I said. "I knew him."

"Well," said Joe, "we took him down to the hospital and there was this real mean head nurse then who was always givin' everybody a hard time, and she decides to throw a tantrum in front of all the other nurses. 'Why the hell are you always bringin' this stuff down here?' she says. 'Why don't you bring it somewhere else?' And Benson, he says, 'Because we love you.' And then she really got pissed off and the bitch crossed her arms and said she wasn't goin' to admit him. Benson, he just walks over to the bag and starts openin' it slowly so that nothin' falls out. The nurse says, 'What's the matter with you, can't you even open a bag?' and she pushes him out of the way; and Benson doesn't say a word about what's inside."

The Rescue stopped short for a red light. We were only a block from the house. My head ached terribly.

"Hey, Joe," I said, "how about laying off the horror stories for a while. You're gonna give me nightmares."

"Hold your horses," he said. "I'm comin' to the funny part now."

I told him I wasn't interested in hearing the funny part.

"Well, Billy is," he said.

"That's OK," said Billy. "I don't care."

"Look," said Joe, "what the hell's goin' on?"

"Nothing, Joe," I told him. "I'm just beat, that's all. Do me a favor, huh? Finish it back at the house."

"Well, I feel like finishin' the goddamn story now," he said. "If you don't wanna hear the goddamn story you don't hafta listen to it."

"Where do you suggest I should go?" I asked him, looking around the inside of the Rescue.

"That's your problem," he said with a shrug. He turned

back to Billy and continued speaking softly, but I didn't listen.

We didn't have any more calls during the night, but I slept poorly anyhow. It's impossible to sleep well at the house; part of you is always awake, listening, waiting. Then, too, my head was hurting and my body ached from dragging the blanket the length of the train.

When I got home I lay in bed all morning reading. After lunch I managed to sleep a little, but when I sat down to supper I felt as groggy as if I'd just been through a ten-round bout.

" 'Yeah' is no answer," said Bev, getting up from the table.

"Why not?" I said, laying down my fork.

"Because when I ask you to guess how much weight Alice lost, I just don't particularly think 'yeah' is much of an answer. That's why not." She started clearing off the plates. "Is that all you're going to eat?"

"Yeah," I said. "Thanks," I added. "It was very good."

She started fixing the coffee. The house was quiet; Larry was still taking his nap and Helen was doing her homework. Her mongoose looked sourly down at me from the shelf over the table. Larry kept a plate in front of the "goose man" and left him scraps of food. Bev would take them away at night; she said all goose men ate at night. I picked a crust of bread off the table and put it on the goose man's plate, but he didn't seem to think much of my offering.

"Maybe if you told me more," said Bev. She was standing at the sink with her back to me.

"What difference would that make?"

"It might make it easier on you."

"Then there'd just be two of us thinking about it," I said.

"Listen," she said, wheeling around, "if you think you're doing me some big favor the way things are now, forget it. This morning Alice said to me, 'When's the last

time you and Larry got away for a couple of days alone?' I
said we'd been too busy with the kids and all. She said, 'Well,
how come Larry's been getting so serious?' She says she
thinks the work is changing your personality."

"What's changing my personality is having her for a
sister-in-law," I said. Then I smiled. Bev just looked at me.
"OK," I said. "All right. Is that what you think, too?"

"As a matter of fact, no. I told her you were just getting
older and settled, that's all. I said that you have more
responsibilities now."

"Maybe Alice is right," I said.

"Well then, why don't you get off the Rescue if that's
the way you feel?"

"And do what?"

"Go back to an Engine Company."

"Nah."

"Then go back to selling."

"I'd hate it."

"Then what about a Ladder Company?"

"I'm afraid of heights."

"Don't put me off, Larry," she said. She poured the
coffee and sat down opposite me at the table. "What are you
getting out of this?" she said.

"I don't know," I said. "Honest to God, I don't know.
Sometimes I get so involved with some of the things, it's as if
they were happening to me. Some of them make me sad,
others just make me sick. But whenever I think it'd be better
to get off, I start thinking how this is the only job that's ever
mattered to me. I start thinking about what else I could
possibly do, and nothing is as important. Whenever I think
seriously about leaving, it scares me even more than staying.
When I was a kid and there was something wrong they'd put
me in my room and close the door. I hated what was going on
out there, but being locked away where I couldn't do anything about it was even worse."

I fiddled with my coffee cup. I found another crumb and fed it to the goose man.

Bev said, "Remember when you first went on the Fire Department? I never thought you'd last a month. You were going to be an executive and you looked so good in a white shirt and tie, and now you came home black and filthy with soot. But you came home so much better than you ever did on sales. You were like a new person."

"Christ," I said, getting up from the table, "if the job stinks, it stinks. No one's breaking my arm."

"It's completely up to you," she said. "You know I'm willing to do whatever you want."

"Yeah, well," I said, "one of these days I'll put in for a transfer and get off."

I left home early and parked near the firehouse. Then I walked through Harvard Square, up past the Old Burying Ground, where the tombstones of the early Puritan settlers jut from the frozen earth, to the Cambridge Common. The colonists grazed their cows there and voted there, and during the Revolutionary War the minutemen met there, before the Battle of Bunker Hill; and there also George Washington took command of the Continental Army.

Now night was falling; just a few traces of red still showed in the west. I walked over to where the kids were playing ice hockey on the frozen ball field. They rumbled over the rough surface on their skates, their sticks hitting the puck with a sharp crack as they swished back and forth, red-cheeked and shouting at each other. An Irish Setter, slipping and scrambling to his feet, circled madly after one of the skaters. Then he lay down next to a small kid who stood in the snow by himself shooting an imaginary puck into an imaginary goal and raising his arms in triumph.

As I walked back toward the house I thought, I've got to separate myself from what happens. I've got to try to make it so that whatever happens has nothing to do with me. If I could only do that, I thought, I'd be all right.

At seven-thirty we had an old man with severe indigestion up near Inman Square. He was a thin old man with bushy white eyebrows and mustache. He was wearing a Red Sox baseball cap that he removed when we arrived. He told us he had high blood pressure and that he'd gone off his diet and eaten some potato chips. "I shouldn't have eaten them chips," he said.

When we put him in the chair he sat up straight, like a little boy on his best behavior. But in the truck he said he didn't feel too good and that he was afraid. He winced and pressed a hand to his forehead and I thought he was going to cry. Billy Stone put a hand on his shoulder. "It was but a couple of chips," the old man said, looking up at us, puzzled. He winced again and Joe put the oxygen mask on him.

A little after ten o'clock we had a call for a knife fight at one of the housing projects. We were silent in the truck going down. Finally Billy got up and put on his helmet, then Joe and I did, too. If we had to make our way into the middle of a gang fight we'd feel safer with them on. But before we got there Fire Alarm radioed that police at the scene said the report was false, and we returned to quarters.

Around eleven we had a call to go down to Greasy Village—a section of Cambridge just below Western Ave., a few blocks from where I grew up. The call was for a possible o.d.

We pulled up in front of a big flat-roofed six-family house. There was a white picket fence with a gate that creaked open as we ran into the yard, crunching over the frozen snow. Nobody was waiting at the front door or in the hallway. The apartment was on the first floor. The door was ajar and we pushed it open.

A young woman lay facedown on the carpet in front of

the sofa. She was tall, and her hair was dazzling yellow, the color of corn. I'd seen hair like that before. I froze. Cooper shoved past me. He rolled her over and my heart skipped a beat. It was Christine.

Once, when I was fifteen, in the summer, Christine dared me to squeeze her hand as tight as I could. She was a big, strong girl, and she wrestled me to the ground. In those days she was lean and well tanned and with that hair she always looked as though she was standing in her own sunshine.

Now saliva dribbled from the corner of her mouth. She'd put on weight; her skin was pasty and lined. I opened her eyes: the pupils were normal. She hadn't gone into a coma yet. Her respiration was good. Her neck was warm, her pulse strong and rapid. I shook her and she groaned as I held her. I shook her again.

Dave and Joe were searching for the pills. They didn't have far to look; Joe found the empty bottle on the table beside the couch. There were balls, blocks and stuffed animals scattered around the room, and Cooper found two young kids asleep in one of the bedrooms. A neighbor had appeared at the door, and he asked her to stay with the children.

Billy Stone brought the chair and we carried her out to the Rescue. Cooper radioed ahead. We gave her oxygen as we raced to the hospital. Her eyes were starting to look fixed and I kept shaking her to keep her from going to sleep. She would open her eyes, look up at me for an instant, then drift off.

The door to her place had been ajar and there hadn't been anybody waiting for us; she must have phoned herself. She hadn't really wanted to die. If she could be kept alive, maybe there would be time for her to work out her troubles.

The summer I knew her there hadn't been any troubles.

I was in charge of athletics at a neighborhood park, which meant I was supposed to put up the swings for the little kids. I was a real big shot. I used to put up one swing for myself and the kids would get off it whenever I wanted to sit down. My friends hung around there with me, and every afternoon she'd show up with her friends and they'd sit on the benches. At first I pretended not to notice her. Later I would put up another swing as soon as I saw her coming.

In the evenings we'd go up to Harvard Square. The guys hung out in front of Charley's snack bar, the girls in front of Brigham's ice cream parlor. Before I walked her home I'd buy her a cone and we'd have to lick fast, as the ice cream melted in the summer heat. Then we'd race down Putnam Ave. to her house and sit on the porch and make out, or go inside the hallway and make out. Or we'd go down to the river and make out on the cool grass with the crickets chirping and the lights of Boston shining reflected in the Charles. Then I'd walk her home and we'd make out on her porch. We were always making out. She'd shake me when it was time for me to leave. In the fall we went off to different schools.

"Wake up, Christine," I said, shaking her. "You're gonna be all right." She was too drowsy to hear me now. Her body was limp, her eyes shut tight. She moaned. Her pulse was weaker and she was barely breathing. Joe gave her more oxygen. I don't know what good the oxygen does; the body can only absorb so much of it. But it didn't hurt her, and for us it was better than doing nothing.

I shook her again. She was asleep now. "Wake up, Christine," I said.

"You know her?" Joe asked me.

"A long time ago," I said.

We rushed her into the Emergency Room. Two of the nurses began to undress her while a third brought the stom-

ach tube and a sterile bowl. The intern rolled the suction
machine into place. I left the room before they slid the
length of tube down into her.

From the lounge, where I sat alone, I could hear the
dizzying drill-like whir of the machine; then, suddenly, the
violent, sickening sound of retching.

Sometimes it's just impossible to comprehend everything that's wrapped up in a single call. You see the whole
span of a life compressed. Only a moment ago, it seems, you
were fifteen years old with nothing more on your mind than
what you were going to do that night. And then you're here
and you see what life and time can do to a person. It's just too
big for you.

The machine was whirring and she was gagging on the
tube that was sucking the poison out of her stomach. Then
the machine stopped and she threw up.

Dave came to the door. "You gonna talk to her?" he
said.

"No. What ever gave you that idea?"

"Joe said you might want to."

"No. She wouldn't know me anyway."

"Yeah, well," said Dave, "there's no rush. Billy says
we're gonna hang around a bit more anyhow."

He left me sitting alone in the room.

I thought how when a guy falls down the stairs and
breaks his leg and you put on a traction splint and do a good
job, you spare him some pain and the risk of a more serious
injury. You've done something for that man, and there's
satisfaction in knowing that.

But suicide's too big. You transport them to a hospital,
they suction them out, and the medical problem is over. But
the psychological problem remains. She had messed up her
life somehow and you couldn't unmess it. It was wrong to
think you could help. Anything you could say would be just
words; they couldn't wipe out whatever awful thing the

years had done to her. She had small children, and she was so far gone that she was willing to leave them. I couldn't imagine leaving my kids alone for even a day—they were my life. I just wouldn't have anything to say to her. It was too big for me.

I heard coughing. Then there was loud crying, and a voice I didn't recognize was screaming at the nurses.

I went out to the main desk, where Cooper and Dave were joking with the receptionist. "Come on," I said, "when are we getting out of here?"

Dave and Billy looked at each other.

"Don't you feel like saying something to her?" Dave said.

"She didn't ask to see anybody she knows," I said. "It might embarrass her."

"It might do her some good," Dave said.

I shook my head. "Let's go," I said. "I want to get out of this place."

On the truck, returning to the house, Joe and Billy Stone rode up front with Dave and Cooper. They were all laughing about something. When the conversation stopped they glanced back in my direction.

I was sitting alone in the rear of the Rescue surrounded by the masks, the tanks and the stretchers. Things were whirling, spinning, happening much too fast. I couldn't keep track of them or begin to straighten them out. It wasn't a question of squeamishness, I told myself; it wasn't a question of fear. I'd been over those hurdles before. It was foolish to push yourself beyond the point of common sense simply out of fear of appearing afraid or weak. That was just another kind of weakness. But was a feeling of powerlessness a weakness, too? Or was it just something you couldn't do anything about?

Rolling down Cambridge Street, the truck seemed to be boring an endless tunnel through the night. The tunnel was like a long hallway leading past room after room in which people were suffering. The doors were locked and I was alone in the hallway. What was I doing on this damn thing? I thought. I didn't want to see any more. In the beginning I had put up with it because I thought I was learning how to help. But now this truck felt like a prison. I was sick to death of riding it. Seeing Christine had made me realize that I'd been kidding myself. It hadn't made the slightest difference that I was there, not really. It was just too big for me.

When I got back to the house I telephoned Bev. She was alarmed; I'd gotten her out of bed. I said I was sorry, that I'd lost track of the time. I said I wanted to know how the kids were. She said they were fine; they were both asleep. She asked if everything was all right. I said yes, and then I told her that I'd been thinking, and that as soon as possible I was going to get off the Rescue. I asked her if she was pleased, and she said fine, OK, whatever I wanted. I said I'd tell her about it in the morning.

"Larry," she said, "you're sure nothing's wrong?"

I said I was fine and that we'd talk in the morning.

"Larry?"

"What?"

"Be careful."

At five in the morning we had a call for a refrigerator leak. Some guy up in North Cambridge had been trying to defrost his freezer by chipping at it with a knife. He was waiting for us on the front stoop, a slight figure in a bathrobe and slippers, shivering in the predawn cold. He said he felt bad making us come all the way out just for a stupid thing like that, but the hissing of the escaping gas scared him. He said his wife used to tell him never to do that.

Cooper told him those things happened, and that it was just as well to play it safe. As soon as we stepped inside we knew it was all right. Only very old refrigerators that run on sulphur dioxide are dangerous, and there's no mistaking the smell when that toxic gas leaks out.

The old man's refrigerator was safe enough. It was an ancient General Electric with a rounded top. The old man watched carefully as Dave and Joe jockeyed it away from the wall. The door swung open, revealing a pint bottle of milk, one tomato and a few eggs.

"What are you doing?" the old man asked as Joe squeezed behind the refrigerator.

Cooper told him that we were just going to crimp the back so that more gas wouldn't leak out. Billy explained that the refrigerator was of no use to him anymore, since it would cost much more to fix it than to buy another used one.

"But this was a new one," said the man, his shoulders as rounded as the refrigerator itself. "I mean, I bought it new. It was a present for my wife, you know."

The old man slumped down into a chair. It was freezing in the kitchen with the windows wide open. "She was always telling me not to do that," he said. "It's just that early in the morning, when I can't sleep, I look for things to do. You know how it is." He looked around the room. "I try to keep the place up the way she liked it," he said, "but if she were to come back now, she'd have a few words for me."

The old man offered to make us tea, but Billy said we had to be getting back. Then the old man asked if it was safe to keep the refrigerator inside. Billy said there was no danger, but you could tell the old man was worried about it. We don't make a practice of this, but Cooper offered to take it down for him.

We got the refrigerator dolly from the truck and carried the thing out. It wasn't nearly as heavy or as awkward to move as the newer ones, but it was badly rusted on the back and bottom, and pieces of it broke away. The old man put on a hat and coat and followed us out and watched us lay it like a coffin on the snow beside the curb. Then Joe ripped off the door with a bar so that any young kids who might play with it couldn't get trapped inside. The old man looked on and Cooper asked him if he was sure he felt all right. He thanked us and said he was fine and not to worry, but when we pulled out he was still standing there in his hat, coat and slippers, his thin stream of breath showing white in the dark, stone-cold air.

It's strange; just when you think you've had all you can take, when you start to hate the very sound of an alarm and dread entering each new house for fear of what you'll find there, just when it all seems futile, some little thing happens that makes you wish only that you could reach out and do more.

In the morning, at breakfast, Bev said she was glad that I had decided to get off. She said she thought the work had become too much for me. Then she looked at me and asked if I'd been serious on the phone.

I said that now I wasn't so sure I wanted to quit.

She asked me why. When I didn't say anything she said I probably didn't feel like talking about it then.

"No," I said. "It's just that I don't know what to say."

"Last night you said we'd talk about it," she said.

"I can't leave," I said.

"Why not?"

"I don't know why," I said. "I wish I did, but I don't. I just can't is all."

=FIVE

IT WAS AN UNSEASONABLY WARM MARCH MORN-
ing. Suddenly, after months of cold, gray weather, the sun
was pouring down out of a brilliant blue sky. Joking and
grinning like fools, we worked in shirt-sleeves out on the
apron in front of the firehouse. The Aerial Company was
drilling on raising the platform, and we were washing down
the Rescue. An unbroken stream of students cut across the
apron on their way to classes, some of them stopping to
watch the Aerial Tower. Bells tolled in the Yard, and a pair
of sea gulls wheeled overhead crying loudly. The change was
fantastic. Judging by the way the seasons were flashing by,
the Commonwealth of Massachusetts had to be spinning
around the sun much faster than usual. It was no wonder we
all felt a little dizzy.

Davey saw her first. All he said was "Uh-oh." Jack
Dillon and Kilroy were standing near us and all heads turned
in her direction. She was crossing Cambridge Street from
Memorial Hall and heading straight toward us. She was

long-legged as a racehorse and wearing a skirt that was barely more than a waistband. There may be worse places in this town for a woman to walk past than our house, but I can't think of any offhand.

I braced myself for a volley of remarks, but no one said a word. The girl looked nicer with each step; I saw just a trace of a smile at the corners of her mouth as she passed. Then we were watching her go: the sunlight on her hair, skirt and legs. The traffic rushing out of the underpass onto Broadway screeched to a halt. A bus driver tipped his cap and waved her across the street. Then she was gone behind the wall of traffic.

"I think I'm gonna faint," said Dave, clutching his heart. He slumped down onto the back step of the apparatus.

"Call the Rescue," said Jerry Martin.

Kilroy bent over and put a hand on Dave's chest. "Jesus," he said, "I can't find his pulse."

"It's lower down," Finnegan said.

Kilroy grabbed Dave's belt; Dave hollered and doubled over.

"You see that, Jack," said Kilroy. "She's killed him. One of the best men on the Department. They shouldn't allow broads to run around like that."

"It's all in your head, Eddie," Jack Dillon told him. "If you respect women as people they can't get under your skin that way."

Kilroy looked at Dillon as if he had a loose screw. "What're you, goin' crazy, Jack?" he said. "Next thing I know you're gonna be advocatin' women on the Department."

"They'd never survive the first night," said Dave, raising up his head.

"You're just gettin' old, Jack," said Finnegan.

"Maybe so," said Jack Dillon, "but I still say that Eddie's problem is all in his head."

"I may be dumb as a donkey," said Kilroy, "but I know that the problem I got ain't in my head. In fact I can show it to you if any of you gentlemen would care to see it."

"Is that the same problem you showed us before roll call?" said Dave.

"Yeah," said Joe Finnegan, "and the same one he showed us after."

"We've all seen your problem," said Dave, getting to his feet, "and we've concluded that if that was our problem we would in fact all have a problem."

"We all *are* going to have a problem," said Cooper, coming around the side of the Rescue, "if we don't get this truck cleaned up."

"That's just what I was tellin' your men here, Guv," said Kilroy. "Knock it off now and get back to work."

We knocked it off and got back to work.

At ten-thirty we had a call for an accident just up Mass. Ave. in front of the law school. As we came out of the underpass and curved to the right we saw the crowd standing in the street. Traffic was bumper-to-bumper, the cars creeping one by one past the crowd, the drivers craning their necks to look.

We jumped off the Rescue and pushed our way through the people. A little boy—no more than five or six—lay on his back in the gutter. His face was turned to one side and there were clumps of blood as thick as custard on the pavement around his head. One leg was folded under the other. Somebody had thrown a sports jacket over him.

Dave and Joe pushed the people back as Cooper and I knelt over the boy. He was conscious, squinting into the bright sunlight. There were tears in his eyes. My heart sank when I saw the blood coming out of his ear. Oh, Jesus, I thought, he's got brain damage. But Cooper pointed out the gash on top of his head; the blood was running down into the ear, not out of it. I checked his eyes. If he had a concussion, the first indication would be one pupil larger than the other; but they were normal. I threw aside the jacket and felt his body for fractures. As I ran my hands down his legs there wasn't a peep out of him. "That's a good boy," I said. Each little foot felt as light as an egg in my palm.

Cooper held a four-by-four gauze against the boy's head to stop the bleeding. I looked up to see if there was anyone with him; people were jabbering all around us. "That sonofabitch took off," I heard someone say. "The bastard just hit him and drove off."

Cooper sent Joe for the orthopedic and held the gauze in place while I tied it up with a triangular bandage. Joe and Billy Stone set down the halves of the orthopedic stretcher on either side of the boy. As the five of us bent over him he looked up, startled, and began to scream for his mother. He tried to squirm away and we had to hold him down.

"Let me go!" he cried. Billy told him we were friends and we were going to help him, because he was a good boy, and we wanted him to lie still; but the kid fought so hard it took all five of us to hold him down, slide the stretcher under him and secure his head with the cervical collar. As we strapped him in, Cooper told him we were going to take him for a ride on a fire truck. "I don't wanna go," he cried. "I want my mommy."

We picked him up and as we were carrying him to the Rescue a woman came running out of nowhere, shouting hysterically. She clawed at Dave, and he grabbed her by the arms and held her off as she tried to fight her way past him.

"What are you doing?" she screamed, her eyes wild, tears streaming down her cheeks. "What are you doing to him?"

"It's OK," Dave told her, releasing his grip on her arms. "It's OK. He's had an accident—a car hit him—but he's OK. We're taking him to the hospital."

"Oh, my God, no," she shrieked, throwing her head back and clamping her hands over her ears.

Dave told her to get on the Rescue with us. She shook her head; he had to lead her up the steps and sit her down on the bench opposite her son. Then we pulled out.

Joe and I stood over the boy, who was quiet now, as we raced with the siren wailing toward Mt. Auburn Hospital. The boy's eyes started to roll back and his lids closed halfway down; he seemed to be sinking into semiconsciousness. I shook him. "Here's your mother," I said. "Look. Your mother's here." His eyes opened wide and when he saw her he cried out to her and strained against the straps. His mother reached out for him but Dave grabbed her, telling her it was best not to touch him, and eased her back onto the bench.

"Is he all right? Just tell me that he's all right," she sobbed. She was young, probably in her late twenties, and her face was all streaked and shiny with tears.

"Sure he's all right," said Dave. "Sure. That's what I'm telling you. It's just a nasty cut is all."

"But there's so much blood," she cried.

Dave told her it was always that way with head wounds. The head bleeds profusely. Head wounds always look worse than they are; lots of times you see people's faces streaming with blood but when you clean them up it's nothing at all. Dave told her all this, but he didn't tell her that no one could know for sure at this point.

"I only left him for a minute," she cried. "I don't know how it happened." She buried her head against Dave's shoulder and cried so hard her whole body shook.

"It wasn't my fault," she sobbed. "It wasn't . . ."

Dave told her no, it wasn't her fault. The boy was quiet again and his eyes were starting to roll back. A sliver of white was all that showed between his narrowing lids. I shook him; he looked up, startled.

"What's your name?" Joe asked, bending over him.

"Uh . . . Mike," he said; his voice sounded very far away.

"OK, Mike," said Joe. "Say, Mike. Hey, Mike. Didn't you ever wanna ride on a fire engine? You're on a fire engine now, Mike."

The boy stared up at him vacantly. Joe pulled a helmet off a hook. He wanted to be doing something, but what can you do? Hit and runs are always upsetting, but it's worst when a kid's the victim. It takes a real bastard to hit a kid and leave him lying in the street.

"Hey, Mike, look here," said Joe. "A fireman's helmet. Wouldja like a helmet? Well, we're gonna give you a helmet. Then you can tell all your friends you rode on a fire engine."

The vacant look spread over Mike's face. He didn't care about helmets or trucks now. He didn't care where he was or what we were. He didn't want anything but his mother. When I shook him I could see the startled panic in his face. All he knew was he had to get to his mother. When you're six years old and you're hurt and scared that's your whole world. "Here's your mother," I kept saying, and those words kept him awake. "Here's your mother," I said, shaking him gently. All the way to the hospital I kept saying it. "Here's your mother. She's here. She's right here with you."

When I was eight years old they took me to the hospital for an infected ear. I didn't want to be left without my mother. I'd always suffered from asthma, and I hated the sight of doctors and hospitals. When the doctor said I had to stay I begged my mother not to leave me, and promised I'd always be good. She told me I always had been good. Then the nurse reached for me and I ran. I ran through the hospital with all the nurses chasing me until I came to a set of doors too heavy to open. When I tried again to get away they strapped me down.

My mother came to visit every day. She said they would only let her come as long as I didn't cry or try to run away.

When the doors opened for visiting hours she was the first one inside, and when it was time to go she stayed until the nurses had to come around and make her leave. She disliked the nurses, because they were cold, and she made fun of the doctors with their stethoscopes, as solemn as men in black ties at a wake. She brought me candy and comic books and fruit that my grandfather sent. She brought me a Monopoly set and taught me to play. She said being in the hospital was like having to go to jail in Monopoly. It was a temporary thing and if you didn't roll doubles on your first three turns, you got out automatically on the fourth. She said it was harder to get out of the hospital by rolling dice, but that two doubles in a row would do it. Every day before she left I'd shake the red dice until they were clicking good and then let them fly across the bed, but I never got two doubles.

When it was time for her to leave she would remind me that if I cried they wouldn't let her come back. And however much I felt like crying, I didn't.

After we took the boy down to Mt. Auburn, we drove back through the Square. I rode up front with Cooper and Billy Stone, and I could feel the hot sun through the windshield. The sign above the bank flashed the time, 12:02, and the temperature, 72. It looked like a holiday in the Square with literally everybody out: the hawkers of underground newspapers, the hippies and guitarists, the vendors of hot pretzels and frankfurters and crepes, the petition pushers, the proselytizers and the Hare Krishna people with their orange robes and shaven heads and bells.

Everywhere on the sidewalks crowds of students and men on their lunch breaks stood watching the girls. I had forgotten how dazzling the girls could be when they took off their heavy winter coats and strolled around liking the feel of the sunshine on their skin and hair. There are as many beautiful girls per square foot here as anywhere in the world.

They give the place part of its wacky magic. You never know what you'll see in the Square. Once in the winter I saw a guy skiing down Mass. Ave. with a Vermont license plate strapped to his back. Once in the summer I saw a band of students dressed in leopard skins and armed with brooms surround and attack a stalled MBTA bus. It's a very special place, Harvard Square; and now the sun was bringing it back to life.

Shortly after we got back to the house Davey went up to make cheeseburgers for lunch. He asked me to see if Cooper wanted any. Billy was at his desk in the officers' dormitory, studying hard for the Captain's exam that was only a few weeks away. The window was open and the breeze rustled the pages of the books spread out before him.

The Captain's exam, unlike the Lieutenant's, is essay-style instead of multiple choice. You have to know the material the way you know your prayers. Each page of the textbooks, each word is importan.. You have to know verbatim about every method of extinguishing fires, from a bucket of water to the most sophisticated chemical system. You have to know everything about the characteristics and behavior of fire, about fire apparatus, fire tactics, fire laws, fire hazards and the properties of materials. You have to know about hydraulics and the structure of buildings.

For two years straight Cooper had been studying for this exam evenings and during vacations and days off. Now that he was coming down the homestretch he was up here studying whenever we weren't out on a call. I asked him if he wanted to eat with us. He looked up and swept the hair out of his eyes. He said he'd brought a sandwich; he couldn't afford to start shooting the breeze with us.

I sat at the kitchen table smoking while Dave, whistling away, made up a tray of hamburgers. He enjoyed cooking and he was good at it. He told me he thought he might have qualified for a loan with which to build his place in New

Hampshire. He said he was glad he had a job with such good security.

"My brother keeps telling me I'm crazy," said Dave. "He always says to me, 'Why don't you get a real job? Why don't you make real money?' But I can't stand the rat race, Lar, and I don't like hustling. I don't want to kill myself making a killing, you know. All I want is enough to live on and the time to enjoy it," he said, centering a slice of cheese on each hamburger patty.

The loudspeaker came on. Billy Stone announced: "*Larry Ferazani. Visitor. First Floor.*"

"Who're you expectin' ?" Dave asked.

"No one," I said, getting up.

"Well, if she's good-looking," Dave yelled after me, "find out if she's got a friend, will ya?"

I went down to the apparatus floor, and it actually was a girl. Deputy Simmons introduced her to Joe and me. She wanted to speak with us about the Rescue. We get a lot of reporters and even an occasional TV crew coming down for interviews. Cooper and Dave both had solid excuses for getting out of this one. Billy Stone wouldn't have minded doing it, but he was on patrol. That left Finnegan and me.

This girl wasn't a reporter; she was doing her master's thesis in some kind of psychology. She wore an expensive-looking suit and she had long brown hair and nice legs. Her name was Marcia. She had a habit of averting her eyes whenever you looked her in the face, and she kept tossing her head as if her hair was always in her way. She wasn't a bad-looking girl at all, but she could have been the Amazon Creature itself as far as Joe was concerned: he knew his work and he loved any excuse to talk about the Rescue.

Joe led the girl over to the driver's side of the Rescue. "Well," he said, laying one hand on the open door and rolling back and forth on the balls of his feet, "here she is. We take a lotta flak from other companies, but I was on an

Engine Company once where we had a brand-new fifty-eight-thousand-dollar pumper—custom-built, which is the only way they come—so I know good pieces. This Rescue is no toy truck. It's a Mack diesel, four hundred horses, power steering, air brakes, dual starters. We've got ten outer compartments where we carry, among other things, a twenty-five-ton MBTA jack, a Porto-Ferguessen hydraulic tool, a high-power circular saw, acetylene torches, seat-belt cutters, glass block smashers, rubber suits for industrial refrigeration leaks, your explosometer for gas leaks, a gasoline-powered generator and then of course your tool chests, lifelines, ropes, grappling hooks and body wraps."

Joe paused to catch his breath and looked up into the cab as though for the first time. "Go ahead," he said, slapping the driver's seat. "You wanna try it out?" He climbed up into the cab.

The girl just stood watching him. "Excuse me," she said, tossing her hair back, "but why are you telling me about the truck? I mean, I don't want to hear about that."

"You don't?" said Joe, his hands on the steering wheel.

The girl shook her head. "Yeah, well, I don't know," said Joe climbing down. I could see he was disappointed. "The Dep said you wanted to hear about the Rescue. So I just thought the place to begin was there."

"Not necessarily," said the girl. "If I'm interested in your emotions and your feelings I don't necessarily want to hear about your truck. Do I?" She smiled at him as though he were a slow pupil.

"You mean the apparatus," said Joe, smiling back at her.

"The *what?*" said the girl, wrinkling up her nose.

"That's what we call the Rescue. We refer to it as the Rescue or the apparatus."

"What's the difference?" she said. "A truck's a truck. Why create a mystique around it?"

Joe's face darkened. "I just thought as long as you were

doing a paper you might wanna get your terms correct."

"But I've already told you, I'm not interested in that aspect of it. For another thing, I'm not here to do some phony public relations job for your department. I'm here to find out the real answers."

"OK," said Joe. "Whatever you want." He glanced at me and I could see he was really annoyed with her.

The girl turned her back to him. "Look," she said to me, "isn't there someplace we could sit down and be more informal?"

We went around to the back of the Rescue and the girl sat down on the hard grated metal steps.

"All right," she said. She uncapped a colored pen and opened her notebook. "Do you ever feel afraid?"

"Never," said Joe.

The girl looked up, smiled, then wrote it down.

"How did you feel at your first fire?"

"Terrific," said Joe. "I particularly enjoyed smashing the furniture and breaking the windows." The girl scribbled furiously, mouthing the words as she came to the end of the sentence.

"When you're going down the street with the siren on does that excite you?"

"Oh, boy, does it," Joe said. "I get warm all over. Just talking about it like this gets me excited. In fact," he said, slapping his belly, "you hafta excuse me." Before the girl even finished writing he was heading up the stairs for the cheeseburgers. I was tempted to go with him, but she was just a young girl, really, and I happen to have a soft spot in my heart for young girls.

She looked up and bit her lip. "What was the matter with him?" she asked me.

I shrugged my shoulders.

"Are there many like him?"

"No," I said. "He's one of the best."

She stared at me for a second. "Well, I'd hate to see some of the others," she said. She gave me a quick glance. "How about you?" She checked her notebook. "Do you get a thrill out of the sirens?"

"No, they don't do anything for me. I don't think they do much for anybody."

"But he said . . ."

"He was just pulling your leg," I said. I told her it wasn't the sirens but the situation. When you know you're going to a fire, then the adrenaline starts flowing. Otherwise you get pretty tired of hearing those sirens.

"How did you feel on your first fire?"

"Scared."

"And your last?"

"Scared."

She looked up at me, frowning slightly, her lips pursed in thought. "Then I don't see how you do this, knowing you could get killed."

"Look," I said "all jobs take their toll in one way or another. I don't see how most people can sit at a desk in an office day after day. Wasn't it Thoreau who said something about every man stepping to a different drummer?"

She said that was getting into the realm of subjective opinion and she was interested in the realm of objective fact. Then she looked down at her notebook and asked me how we managed to look at badly injured people. I told her you found out very early whether you could or you couldn't. It was the same for us as for people who worked in hospitals; you had to put aside your own reactions to be of any help to anybody.

"I mean I don't see how you can do it," she said, "without being a little . . ."

"Wacky?"

She smiled. "I would have used another term."

"What's the difference what you call it?" I said. "If you

come in here with a bunch of preconceptions there's really nothing I can tell you, is there?"

"But I didn't," she said, tossing her hair back. "Listen, this isn't a public relations job. I'm doing research."

"I know," I said, "but while you think you're getting the real answers all you're actually doing is repeating every single stereotype in the book about firemen. Each guy here has his own reasons for doing what he does, most of them very simple and practical ones—like their fathers did it, or they like the challenge, or the security. Maybe they couldn't find anything better, or maybe this is what they wanted to do all their lives. Some of the reasons may be a little deeper or more complex. But whatever their reasons, each of the guys also feels he's making a contribution here that he couldn't make elsewhere."

"All right," she said, "but what makes it possible for you to do it emotionally? I mean you personally. I don't understand that."

"I don't either. You'd probably have to ride the Rescue with us for a year to find out. And you'd have to live with us."

"I don't think my husband would appreciate that," she said, smiling.

"Maybe he could do some research of his own on the Emergency Room nurses while you were with us."

"He might go for that," she said.

"Not when he sees the nurses," I said.

She stood up suddenly, smoothing out her skirt. "I think I'd better go now," she said. As I showed her to the door she said, "I still don't understand why you do it."

"If I knew the answer to that," I told her, "I probably wouldn't be here."

I went upstairs for lunch. Joe was sitting at the end of the table, staring moodily at his half-eaten cheeseburger. I sat down beside him and started eating.

"What the hell was wrong with that broad?" he said after a while. "They come in here and act like you're some kind of freak."

"She was just young," I said.

"She was a pain in the ass, that broad," he said. "Whadda they bother comin' for if they already know all the answers? You didn't give her nothin', did you?"

"No," I told him, "I don't think she got anything from me."

After lunch we had a long run down to Second Street in East Cambridge for a false alarm. As we were driving back to quarters some guy ran out into the street and flagged us down. He said there was a woman lying in the alley; he waved in the general direction of an alley about thirty yards away and then walked off the other way.

While Cooper notified Fire Alarm we ran into the alley. The woman lay on her back, her arms slapping uselessly at the ground like the flippers of an overturned sea turtle. There was a grocery bag on the ground beside her. She wasn't much over thirty, but you could tell at a glance that she'd had a hard life. She had on a huge gray cloth coat that was stained in spots and missing buttons. She was very thin; her calves, showing bare between her slacks and loafers, looked about as big around as my wrists. There were a few clouds overhead now, and in the shade in the alley it was damp and almost chilly. There was something about this weather I didn't like. It wasn't really what it seemed to be.

"She's a regular," said Joe. "I seen her before. I bet she's half in the bag now."

Dave knelt over her. "What happened, ma'am?" he asked her.

"Whaddaya friggin' think happened?" she shot back. "I was goin' and I tripped on this goddamn sidewalk. I dunno what happened." She slurred her words and her mouth was slack. There's nothing uglier than a woman who's drunk. They're uglier and nastier than a man any day of the week. This one wasn't bombed, though; she was just one step beyond feeling good. She'd probably been staggering a little before ending up flat on her ass.

Dave and Joe examined her for a possible spinal injury. Dave had her squeeze both her hands, then he removed her loafers and made her wriggle her toes. Billy Stone brought the orthopedic stretcher. "Watch out for my fuckin' back," she shouted as we scooped her up. "Why don't the fuckin' city fix these walks?" The sidewalk had a few cracks in it, but it wasn't particularly uneven. Dave told her it was terrible the way people didn't seem to care about each other's welfare any more.

"The sonofabitches haven't fixed this walk since the day it was laid," she said.

Joe picked up her bag and we carried her to the Rescue and started for the hospital. We left the back doors open because of the weather.

"Have ya got a cigarette?" she asked Dave. "I wanna cigarette."

"You can't smoke here," he said.

"Don't shit me, honey. Just gimme a cigarette."

"I'm not kidding you, dearie," said Dave. "There's oxygen here. You can't smoke."

"Oh, you're a real heartless bastard," she said. "You know that? You really are." Suddenly she winced and groaned. "I'm cold," she whined. "I feel cold." I got out a blanket and handed it to Dave and he spread it over her.

"Shut the door," she said. "I feel cold."

"It's seventy degrees," said Dave.

"I'm cold," she said. Joe got up and pulled the doors closed. The woman grabbed my wrist. Her hand was cold and rough and I could feel it trembling. With her other hand she rubbed her face. There was an old scab over one eye and her nose was lined with little red veins. She smelled terrible. She grinned at me, revealing an uneven row of tobacco-stained teeth, and squeezed my wrist hard; I pulled away from her and sat down in the far corner of the Rescue.

"I hope I'm not hurt," she said to Dave. "Do you think I'm hurt?"

"You're all right," he said.

"Where'm I goin'?" she asked. "I gotta get back."

"Cambridge City Hospital," said Dave.

"What the frig?" she snapped. "I ain't goin' to no Cambridge City Hospital. I gotta get back."

Dave told her they'd just check her over and she'd be back in no time.

"I gotta get back *now*," she whined. She wiped her eyes and grabbed Dave's arm. He let her hang onto him. "I was just goin' out . . ." she began, her voice tearful. "I just went down the street to get a bottle of milk and I slipped. I gotta get back 'cause my kids're all alone." Her grocery bag, on the bench beside Joe, was torn down one side. We could all see that inside it there was nothing but two six-packs of beer and some cigarettes.

"I just went for milk," she cried, slurring her words. "They're waitin' for me all alone. I don't wanna go to no Cambridge City Hospital. I never went there before."

She stopped, her lips slack, her mouth hanging open. Suddenly she grabbed at her throat, making a hard choking sound as the vomit gushed from her mouth. The brownish stuff ran down her neck and the front of the blanket. Joe lifted the head end of the stretcher slightly so she wouldn't choke. Dave wiped her mouth clean with a towel, then opened her mouth to make sure it was clear. She started to vomit again. He held a bowl up to her mouth, but she pulled her head away and threw up all over the floor. The Rescue filled with the nauseating stench.

I opened the back doors and stood out on the rear step,

holding onto the handrails for support. I thought about her little kids at home, staring at the door, wondering where she was and scared to death of losing her. Looking back inside, I saw that she was stretched out like a skeleton.

I remembered some of the women in the tenement on the dead-end street in back of the house where I grew up. They, too, were emaciated from a steady diet of booze, and their kids were always hungry. They lived in unheated, cold-water apartments. We used to play on the tenement steps, and the women would tramp back and forth all day long to the liquor store around the corner, some of them wearing sunglasses to hide the black eyes their men had given them. Sometimes we heard them screaming from the beatings, and sometimes we heard the men yelling and pounding the locked doors until they smashed them in.

Sometimes they did other things with their men and when we tried to listen they would throw buckets of filthy water at us. They were very mean and they would yell at us for no reason, and if we ever told a bill collector they were at home they would try to get us when he was gone. But they couldn't run too fast and they were always falling and hurting themselves anyhow. Sometimes the bill collector would wait for them to chase us, and in that way he would catch them.

My mother was never mean like those women. She laughed a lot and we always had great times together. My father was on the road selling and my older brother was always out, so my mother and I spent all our time together. I'd play or do my homework in the kitchen while she worked. When neighbors visited she'd tell them how well I was doing in school and how I was going to be an altar boy, and I'd sit there soaking it up. When they stayed too long we'd make faces behind their backs, and we'd laugh about them later over coffee. My coffee was almost white, with just enough coffee in the milk so I could make believe we were drinking together. We did everything together. When she tucked me in and kissed me good night she would say that old people were no fun, that she only liked young men like me.

When she didn't feel well she would go to bed early. One night when I was about ten I thought I smelled something cooking. I went in the kitchen but there was nothing on the stove. Then I looked down the hall and saw the smoke seeping out from under her door. I ran, but it seemed miles away. The smoke poured out as I threw open the door. "Momma?" I cried. It was pitch-black and the smoke stung my eyes. I felt along the wall for the light switch, but I couldn't find it. I made my way to the bed and touched her. "Momma, wake up!" I shouted, shaking her. It was her mattress that was on fire. "Momma, please." I shook her and slapped her but she wouldn't move. I tried to pick her up but she was too heavy for me. I rolled her off the bed and dragged her out into the hallway. Then I ran for a bucket of water, dumped it on the smoldering mattress, ripped out the burning part and stamped it out.

In the hallway my mother came to. I couldn't stop coughing and my mother was scared that I was having an asthma attack. When she found out that I was safe she hugged me and kissed me and laughed. My eyes smarted, my fingers stung, and I felt sick to my stomach from the smoke, but I was tremendously happy and my mother was laughing. I thought there was nothing in the world that could ever take her away from me.

After taking the drunken woman down to the Cambridge Hospital, we cleaned the truck and went back to the house. Kilroy, Martin and Riordan were laughing it up in the patrol room.

"Hey," shouted Martin, "you guys just missed a visitor from New York."

"Where's that?" said Davey.

"Don't get smart," said Jerry Martin. "This guy gave us a real serious talk on forcible-entry tools. He'd been a locksmith or something before, so he knew his locks. He was demonstrating one technique and he asked us if we had a certain type of door up here in Cambridge. Like he thought this was the sticks and we didn't have doors up here."

"So then," said John Riordan, "the guy asked how many of us understand electricity. No one says anything. So the guy says, 'Oh, come off it, don't you have an electrician around here?' So Eddie says, 'Elect—what?' 'Electrician,' says the guy. 'Yes sir,' says Eddie, 'but only every four years.'"

They all cracked up. We left them there laughing and slapping their legs like idiots, and asking each other whether in Cambridge there were ceilings, cars, wheels, fire.

Billy Stone and I stood out in front of the house for a while. Billy had a head cold or an allergy, he wasn't sure which. He said he'd been up late working on a hutch for his pet rabbit and the sawdust had gotten to him. He was worried about the neighbor's cats and dogs getting at the rabbit. I wondered if a rabbit might make a good pet for Helen and Larry and he started to tell me all about them.

A car pulled onto the apron and rolled up to us: an old Chevy, North Carolina plates, no hubcaps, a squeaking engine. There was a husky guy behind the wheel, beside him a woman with the beginnings of a double chin, and two kids fighting in the back seat.

"Y'all tell me how I git to Harvard University?" the guy asked, leaning out the window.

"You're here," I said. "It's all around you."

He looked back at Memorial Hall across the street and grinned. "Well now, is hit true what they say about this being the oldest university in America?"

I told him it was, that the college was founded in 1636.

"Now that buildin' back there," he said, gesturing toward Memorial Hall. "About how old do you reckon hit is?"

I said I didn't know for sure but that it was dedicated to Harvard students who had died in the Civil War.

"Well, we're real sorry about that, you know," he said. Then he pulled out.

You forget how much history there is in Cambridge when you've lived here all your life. Once on a trip south I drove hours out of my way to visit Colonial Williamsburg and see a bunch of restored buildings set off in the middle of nowhere with costumed actors running around. I had to laugh, thinking of how many times I had walked up Brattle Street right past actual colonial houses that were still actively a part of the life of the town. It was much better to look at the real thing and wonder how it looked hundreds of years ago, than to have history turned into a sideshow.

Two girls pulled up in a convertible with the top down. Billy took this one.

"How do we get to King's Row in Brighton?"

"You go back through Harvard Square," said Billy.

"Oh, do we have to go back through Harvard Square?"

"Well, I could send you another way."

"Is it a good way?"

"Yeah. It's a terrific way," said Billy.

I wish I had a quarter for every person I've given directions to. With its spider's web of narrow, one-way streets, Cambridge is like a maze. If you stand in front of the house for any length of time you get the feeling that most of the people driving around this town are lost.

After an hour we went inside. It was clouding over and getting chilly in the shade. Some days are really slow and nothing much happens.

A little after four o'clock we had a call for a woman trapped in an elevator at a housing project. The last time we'd been to this particular project on this kind of call it had been for a kid who'd lost his balance while playing on the roof of the elevator. The elevator was stuck at the top of the shaft and we had to climb up to the roof of the building and enter through the skylight. The boy must have been about fifteen. He was jammed between the elevator and the wall, with only his head and one arm showing.

That time we used the hydraulic jack to force the elevator off its track away from the wall, holding on to the beams overhead at the top of the shaft in case the whole elevator should drop. At one point a priest appeared in the window above and asked if he could come down and administer last rites. Cooper told him it might be better to wait until we had finished.

"Has the boy been dead long?" the priest asked.

"A few hours, Father," said Cooper.

"Well, I suppose, then, he could wait a little longer."

It turned out to be quite a while. The extrication took us close to two hours. Near the end a cop appeared in the window with a flash camera, wanting to take pictures. Cooper kicked him out.

This time when we pulled up in front of the red brick high-rise building the police told us that a woman and a child were trapped between the third and fourth floors. The woman was in labor. We hustled up the stairs to the fourth floor carrying a stretcher, a halligan bar, a tool kit, sterile sheets and the maternity kit. The hallway was crowded with cops and residents. We could hear the woman's muffled cries from inside the elevator. Cooper got down on his knees and put his ear to the outer door. He asked one of the cops if the power was off.

"Don't know," said the cop with a shrug. Cooper dispatched Swanson to make sure it was off. Then he tapped on the door and asked the woman if she could tell him how many minutes apart the contractions were. She said more than five. Billy told her that was good and not to worry, that we were going to open the doors immediately.

Joe jammed the bar between the outer doors and wedged them open a crack, and Cooper and I pulled them apart. The roof of the elevator came up to our hips. Joe bent down and jammed the bar into the top of the inner doors and Cooper and I got down on our knees and spread these doors open. The woman was standing in the back of the elevator, her hands over her eyes, crying hysterically. Her head was at the level of our ankles. She was a young girl, in her early twenties, and her belly protruded enormously. A small boy, wailing also, clung to her maternity smock. Cooper squeezed down into the elevator, talking to her soothingly, then Joe and I squeezed down. Billy Stone lowered in the stretcher and I handed the screaming child up to him. The woman was still crying hysterically. We stretchered her and Cooper checked to make sure the baby wasn't crowning. Then we strapped her in and covered her with a sheet and the four of us slid her up and out.

One of the neighbors was holding on to the child. She was telling him that his mother was going to be all right and that when she came home she'd bring a new brother or sister for him. The kid didn't seem too impressed with that; he wanted to stay with his mother. But his mother was too upset herself even to notice him.

She began to calm down in the Rescue on the way to the hospital. She stopped shaking and just cried softly to herself. Dave asked her if she felt as though she had to strain or to move her bowels. She said she didn't. He asked her if she had broken her water yet. She said she hadn't. He asked her how she knew she was in labor.

"Are you kidding?" she said, looking up at him. "This is my fourth." She told him she was eight and a half months along and hadn't been expecting it so soon. She had called her husband at work but decided she couldn't wait and phoned for a cab. Dave asked her where her husband was now.

"Anywhere but where I need him when the time comes," she said. "He must be psychic."

Just then the pain hit her. She clutched the sides of the stretcher and threw her head back and screamed. She held on so hard her knuckles went white and sweat broke out on her face. It lasted about ten seconds. When it was over she closed her eyes and breathed deeply in relief.

She seemed much calmer when we reached Mt. Auburn. She'd been through it before and it must have seemed all downhill from here, because she smiled and thanked us. Davey said it was a wonderful thing to be having a kid, and she said yes, it's wonderful, isn't it?

I remembered when my own mother's belly got big and for a long time she made everyone believe she was pregnant. Everybody said how wonderful it was for a woman her age to have another child and I was very glad because it seemed to make her so happy. Once when Mrs. Grinley from next door asked me how I'd feel about having a new brother or sister I said I didn't care. I said I didn't need anyone but my mother, and my mother hugged me so hard I had to pull away.

Her stomach grew bigger all winter. It made her very tired and she couldn't go out and we played a lot of games together indoors. She kept kidding about the way she looked, asking if I minded her all swollen up, if I found her ugly or just the same. As her stomach got bigger she tired more easily and had to spend a lot of time in bed. When her time was near, my father came home from a road trip, and they sent her away for a few days' rest. They said she had to rest extra hard because it had been eleven years since she'd had me. So she went away for a few days.

At the end of the week, when she still hadn't come home, I asked my grandfather to take me to see her. Instead he took me to the rodeo. He said he wanted to talk to me. The rodeo was dull, except for the part where a clown kept trying to set up a fence and his horse kept knocking it down when he wasn't looking. On the way home my grandfather told me that when he first came to this country he was as stupid as the clown. He went to work digging foundations with a pick. It was hard, backbreaking work and the foreman told him the first day that whenever he saw a guy with a shirt and tie on, he'd know that was the boss and he'd have to work harder. It was one of those downtown construction projects where the spectators passed by all day long, and Jesus, my grandfather said, they all wore ties.

Two weeks later they took me to see her. My father brought me over to the Cambridge Hospital, but a nurse led me up to see her alone. She had a room to herself. She looked very tired and her face was different. Her belly was swollen enormously under the blankets. The nurse said I couldn't stay long and when she left I asked my mother why I couldn't come to see her every day. She said they wouldn't like that and then they'd never let her leave. I told her I'd brought her something that would get her out, and I took the red dice out of my pocket. All she had to do was roll two doubles. At first she wouldn't hold them, but I made her. She was too weak to roll them so I put them in her hand and she kind of let them drop. She made a doubles. I quickly picked them up and gave them to her, but she just held them and didn't let them fall.

Then the nurses came to say I had to go. My mother told them to wait outside just a minute more. They went out and closed the door.

My mother lifted herself off the bed and hugged me good-bye. She hugged me so hard she began to tremble. She hugged me as if she would never let go. I didn't want to leave but I didn't see how I could stay. Then the nurses came and pulled her away, and the one that took me back downstairs started to cry.

The next day my father took me to my uncle's. He said I

had to stay there awhile and he left me. I remember sitting in the kitchen when my uncle came out wearing a white shirt; in his hand he held a black tie.

They took me home for the second day of the wake. The weather was unseasonably warm and the windows were open. I waited in the long hallway before they let me see her. I was afraid of how she'd look because they said she'd had leukemia and I didn't know what that meant, but when I saw her I was relieved; the painted face in the casket wasn't hers. I was sure that she would come back, because I could never believe that she was dead.

$=$ *SIX* $=$

"THEY STINK," SAID EDDIE KILROY. "IT'S GONNA be the same old story all over again."

As usual a bunch of us were shooting the breeze outside the main door in the spring twilight. It was early Friday evening; our group was scheduled to work the night shifts both Friday and Saturday. Up in the Fire Alarm room overhead we could hear bells reporting boxes struck in the Boston area, but things were quiet in Cambridge. Groups of kids sauntered past the house on their way to the cafes and theaters of the Square. The streetlamps came on, lighting up the fresh green of the new leaves all around us, and the air was charged with the excitement of the oncoming night.

"I still hafta like their chances," said Dave. "If their pitching holds up, I can't see who can beat 'em."

"Who cares," Kilroy said. "The Sox're a bunch of losers. You couldn't pay me to go down to Fenway Park."

"What if all it cost you was a nickel?" said Jack Dillon. "That's what we used to pay when I was a member of the knothole gang."

"So was I," said Jerry Martin. "We used to go down to Braves Field."

"That's right," Jack said. "You got a card stamped at a city playground and you could go to the game on that day for a nickel, and, Jeez, we saw some good ones. I remember one that lasted sixteen innings. They didn't have lights there and it was almost as dark as now. You could hardly see the ball from the mound to the plate. So Frisch of the Giants, he comes up to bat with a lantern. Those were the days."

"Talk, talk, talk," said Kilroy. "That's all you old guys are good for. Next thing I'm gonna be hearin' about your crossin' on the Mayflower."

"It rained the whole time," said Jerry Martin. "We couldn't get no reception on the wireless."

"I don't remember it raining on that particular voyage, Jerry," said Jack Dillon, "but I do remember that them were good days. I wish I could go back."

"Not me," said Kilroy. "There's no way I want to go through being a kid again. You couldn't pay me to go back. You guys are gettin' senile."

"Jesus," said Jerry Martin, crossing his arms. "You're in a fine mood tonight. You haven't got a good word for anybody."

"That's right," said Kilroy. "I haven't got a good word for nobody." With that he got up and went inside.

"Christ," said Jerry Martin. "What the hell's eatin' him tonight?"

"Kilroy's just a kid," said Jack Dillon with a shrug. "There's no figuring them out."

From inside came the single sharp ring of the phone. Dave ducked in the doorway and emerged a second later clutching a slip of paper. "Rescue going out," he shouted as the bell rang throughout the house. First Joe, then Billy Stone, then Cooper slid down the poles to the apparatus floor. We all ran for the Rescue. The address was Boylston Street. They hadn't said what the trouble was.

In less than a minute we pulled up in front of a small brick building next to a little outdoor cafe. The customers were all on their feet as we jumped off the Rescue. A man came running out of the building waving his arms. "She's in here," he shouted. "Hurry up."

He led us down the stairs and along a semidarkened basement corridor. It looked as though they ran a small book bindery down there. The guy stopped short at the end of the hall. "She's in the bathroom," he said, wiping his forehead. "She just walked in off the street. I never saw her before."

The sign on the door said MEN. Inside, lying on the floor, was a young girl in women's-lib overalls. She was about sixteen, and she was practically bald; clumps of chopped-off hair littered the floor. She stared up at us, shaking convulsively. There was a little pool of vomit beside her on the floor and a thin stream of mucus trickled from her nostrils. A faint gurgling sound came from her throat.

"Look," said the man, pointing to the sink. In the washbowl was a plastic container of drain cleaner. It was empty.

The girl started screaming, and as Cooper knelt over her she twisted around and tried to reach a pair of scissors on the floor nearby. Joe kicked them away. She started hitting Cooper. I dropped to my knees and held down her arms as Cooper pried open her mouth. It was completely purple inside.

"Let's get her the hell out of here," Billy said. She struggled as we lifted her into the chair. We rushed her out to the Rescue, the siren came on and we pulled out fast.

Joe and Billy Stone stood behind the girl, the perspiration streaming down their faces. I stood in front of her with one foot braced against the chair. She kept tossing her head back and gulping as if she was trying to swallow. The stuff must have been burning up her insides. I heard Cooper on the radio reading off the chemical contents of the drain cleaner from the container. In the hospital they had a thick book listing poisons and antidotes. With a strong acid you can't use the kind of antidote that induces vomiting; the last thing you want to do with someone who has swallowed an acid, like this girl, is to flush it back up through her system. They would have to try to dilute it right inside of her. Even if they did manage to save her, every second that stuff remained in her it was eating her up.

The air horn trumpeted furiously. Traffic was at a virtual standstill on Mass. Ave. in both directions. Dave pulled out quickly into the oncoming lane, then cut sharply down Dana Street, sending Joe and me slamming into the side compartments. Billy Stone felt her pulse; he said it was slow. The girl's eyes were on me. She was a heavy girl, and with her smooth complexion and bald scalp she looked like an enormous infant. There wasn't a sign of pain on her face. She chewed anxiously at her lip and mumbled something, swallowing hard and nodding her head. "Time," she whispered hoarsely. Her eyes were on mine. She was trying to tell me something, but I knew it was bad for her to talk.

She took a deep breath and gulped down the air. Her body jerked spasmodically and she raised her head up erect, the muscles in her neck bulging. I could barely hear her words. "I'm gonna die," she said faintly. She tried to clear her throat. She shut her eyes. The cords in her neck stood out as she strained to speak. Her lips moved again, but there was no sound. She opened her eyes wide, as if suddenly surprised. Then she just looked at us in silence.

We rushed her into the trauma room, where the emergency staff was assembled. The doctor asked her her name. Joe told him her voice was gone. As we lifted her onto the table she closed her eyes. They shut the door behind us when we left the room.

We returned to the house. The TV was blaring in the patrol room, where they were all waiting for us. Harry Mondello clicked off the set. "We heard over the radio," he said. "How bad is she?"

"Bad," said Cooper.

"Did they find out who she was?" Jack Dillon asked.

"No," Billy told him. "All they know is that this wasn't the first time she tried it. There were scars on her wrists." Billy put his fingers to his cheek. The girl had scratched him good in the men's room; there was a long red line running down his face. He picked up a scrap of paper from the floor, crumpled it into a ball and threw it into the trash barrel. "Drain cleaner," he said in disgust, shaking his head. "For Chrissake."

No one said anything for a while, then finally Jerry Martin said, "Well, you lost another one."

"That's right," said Kilroy. "You guys're makin' us look bad."

"You're lucky this isn't a ball club," said Jack Dillon. "With your record you'd have been sent back down to the minors long ago."

"Or traded to New York," said Jerry.

"No. Even New York wouldn't fall for that," said Kilroy.

"OK," said Cooper, "I get the message. There's not much point in hanging around where we're not appreciated."

We trooped upstairs and sat at the long kitchen table drinking coffee. No one said anything more about it. What was there to say? Pressures build up, someone retreats into his own small room. There's a big world right outside the door, but they don't know it. All they know is that room where the walls keep pressing in. There are national studies that indicate that suicide ranks as the second leading cause of death—after accidents—for people between the ages of

eighteen and twenty-five. And how many of those automobile accidents, falls, drownings and overdoses of drugs are
also suicides?

You understand how it happens, but you don't know
why. However serious her problems were, other people
coped with similar things and survived them. Why did this
girl choose to die? Maybe there wasn't any reason. Maybe it
wasn't a question of choice at all, but something chemical in
the cells or the blood. In the end a person simply had to want
to survive. And maybe that desire just wasn't in some of us.

A quarter of an hour later bells were coming in. Box
4731, Harvard University dormitories. Everything was going.

The horns and sirens echoed off the buildings as Dave
navigated his way through the traffic in the Square. Night
had fallen and crowds lined the sidewalks to watch us pass,
their hands over their ears to ward off the ear-splitting wail
of the sirens. The last time I'd been to a fire at these dorms
Billy Stone had had to pull me out. We'd had the old All-
Service masks then. Billy and I were crawling blind down a
smoke-filled corridor looking for the room the fire was in,
when my mask clogged. I sucked in hard on it, but nothing
came. Then I lunged in the dark for Billy, grabbing him by
the shoulder. I thought my lungs would burst. He knew
instantly what it was and ran me back down the corridor,
smashed out a window and pushed my head out into the
sweet air.

This time, though, there wasn't any smoke showing. The
apparatus lined the stately driveway in front of Winthrop
House, their lights flashing, their engines throbbing. A huge
crowd of students was milling around on the lawn in front of
the building, cheering, catcalling, applauding as we climbed
down from the pieces. In front of the main doors the Deputy
was talking to the housemaster and the custodians; a film had

been disrupted by the false alarm. We stood in the middle of the grinning students in our long coats and helmets.

"Hey, you dudes," someone yelled. "Lay down your axes and stay for the flick."

A girl shouted, "Take me to your leader!"

They all laughed and then somebody did a pretty fair imitation of a siren.

"Mr. Fireman," someone shouted in a falsetto from the rear of the crowd, "can I see your hose, please?"

Most nights we would have joked back, but tonight none of us felt like it. As we climbed back onto the apparatus they cheered and made the peace sign with their fingers.

"Fuckin' kids," said Joe.

"Hell," I said, "if you were a kid you'd be out there with them."

"Nah," he said. "I didn't do that kind of stuff when I was a kid."

"I meant if you were a kid now."

"I couldn't be that kind of a kid," said Joe. "This reminds me of April, '69."

It did feel a little like that April, the month of demonstrations, strikes and mass meetings, when the students occupied the Harvard administrative offices and the police were mobilized to bust them. Over four hundred helmeted riot police had stormed the building at dawn, and we had been called in to treat a Radcliffe girl who had injured her back when she panicked and jumped from a second-story window. I still remember how the Yard looked in the pale early light, filled with over a thousand students chanting in outrage as the police hauled off busloads of protesters.

One night the Tactical Police Squad, on alert for a possible riot, used our firehouse as a control center. Most of us weren't too happy about that. I knew a couple of the cops from grade school, and while a few seemed to be hoping for trouble, most of them found the whole thing distasteful. I didn't think the students should have taken over University Hall, but they hadn't started the war and they did have a right to protest and to be heard. It seemed incredible that people should be preparing to fight one another here in Cambridge over a situation created elsewhere by the federal government.

But there was no riot then. There was one a year later, in the early summer, when the Square had filled up with the usual transients, and it was more an act of vandalism than a gesture of protest. It transformed the Square into a war zone. Barricades were erected, rocks thrown through store windows, cars overturned and burned, buildings set ablaze. Firemen rushing in to extinguish the flames were jeered at and threatened and stoned. Police in full riot gear lobbed tear-gas grenades into the crowds; the wailing of sirens and the acrid fumes filled the night. The "trashing" of Harvard Square cost the students much of the support they had earned in the community.

When I was growing up, relations between townies and students were bitter. We looked upon the Harvard kids as different animals. They were weird creatures who lived in a fog, crossed streets without even looking, rode English bicycles, drove small foreign cars, wore crazy clothes. They were all rich, but they were tightfisted. I remember trying to pick up spare change shining shoes after school in Harvard Square. Guys who worked Central Square said I was making a mistake. They were right. You told a student the price was a dime and that's what you got, a dime. It was better up at Central Square, where the ordinary people, at least, would tip you.

Most of our feelings were pure resentment. Those kids were studying at one of the greatest universities in the world, and growing up in the shadow of Harvard we could never forget that. Everything inside the grassy Yard was beautiful: the red-brick colonial buildings, the big library with its granite steps and columns, the white spire of the church. Sometimes we peeked through windows into rooms with chandeliers, mahogany walls and men smoking pipes in sturdy leather chairs. It was a magical place that led in ways we didn't understand far beyond the boundaries of Cambridge; while we, with difficulty, were trying to scrape up carfare for a trip into Boston. We were paupers on the palace grounds, and when they locked the gates at night we always found ourselves on the outside. For all the years that we lived in sight of its domes and spires, Harvard taught us one lesson: that it was hallowed ground and we had no place in it.

Like a band of Indians we raided this richest of universities. We roamed over the Soldiers Field athletic complex gathering old tennis balls for our stickball games, and hunted in the thick grass on the edge of the lacrosse field for the hard rubber balls that were lost there. We occupied the Business School baseball diamond, swam secretly in the indoor pools, slipped into the best seats at football games and skated for free on Watson Rink. We were delighted with ourselves for outsmarting the geniuses.

Compared with the people at Harvard we were poor; but then, so was everyone else where we came from. Our heavy shoes soaked up the water and cold like cardboard, our cheap woolen pants itched terribly, and in school the lice jumped from head to head. When we got to scratching like dogs we gave ourselves treatments at home, spreading on the stinking gook at night, covering it with a towel and washing it out in the morning. In a few weeks we had them again.

For fun we fought each other for control of our respective corners, swam in the polluted Charles, chased the girls on the banks by the Weeks Bridge and slept out under the stars. It never seemed to me that it was necessary to be rich to have fun in life.

But now almost all the old family neighborhoods are gone. The tremendous growth of Harvard and MIT from opposite ends of town has driven up rents and taxes, and chased the middle-income working people out. Increasingly, Cambridge has become a city of the young transients and the elderly, of the very rich and the very poor.

When we got back to the house we settled in the patrol room, drinking tonics and watching the ball game. The Sox were down by a run with the bases loaded in the seventh, none out, Smith at the plate and Yastrzemski on deck. As the Yankee pitcher looked in for the signal, Kilroy, on the pretense of fixing the focus, switched the channel. He was persuaded to switch it back, just in time for us to see Smith foul out. Then Yaz ripped the first pitch into a second-to-short-to-first double play.

Half a minute later the phone rang. Mondello took the call: a man threatening to jump out of a window on Franklin Street.

"Jesus," said Dave as we got to our feet, "the guy could at least have waited till the end of the ninth."

We went out along Mt. Auburn Street, turning down Putnam Ave. at Kerry Corner and running right past my old house. In the old days I would have known each of the faces flashing by. Now I didn't recognize a soul.

We turned left onto Franklin Street, pulling to a stop just past Hancock, in front of a three-story brick tenement. A huge shirt-sleeve crowd stood gazing up at a second-floor window where a guy in a T-shirt and a white football helmet was leaning out. Both of his arms were dripping with blood and there was blood smeared all over the helmet, which flashed red in the light of the Rescue. "Water," he moaned. "Water."

It looked to me as though he'd been shot. Someone shouted that there was a girl with him. I heard the distant wailing of police sirens as we moved to the door directly beneath the window. Something—blood—splattered hard on my shoulder.

We tramped up the wooden staircase. My heart was hammering. There was no telling what had happened up there. Some maniac could be waiting to blow our heads off.

The door was unlocked; Davey threw it open.

The guy backed out of the window and, turning slowly around, straightened up to his full height. He was enormous: six feet six, two hundred and forty pounds at least. He took one heavy step toward us. In the semidarkness we couldn't see his features; it was like being face-to-face with a monster. The apartment was a shambles of toppled lamps, smashed chairs, torn books, shattered records and plates. He took another step toward us, broken glass crunching under his boots. As we played our lights quickly over him he shielded his eyes with one bloodied hand and swatted at the beams with the other. Blood was running from his arms and wrists. "Take the lights off him," shouted Cooper, moving toward him.

"Let's look after those cuts," said Billy. We all advanced toward him. The guy just stood there, watching us approach, his hands curled in front of him as though he were holding an invisible ball. We knew we could subdue him if we had to, but he was huge and there was no telling what he was going to do. I'd been in plenty of jams before but this was different. In a jam, you know what's coming. Here we had to try to restrain him without alarming him or causing him any physical harm. It was like stepping into a bullring with a cube of sugar.

Cooper halted in front of the guy and held out his hand. "Your arms are cut," he said. "Let us take care of them."

The guy looked down at Billy's hand as Davey, Joe and I spread out to surround him.

"That's good," said Cooper. "Everything's gonna be all right." Cooper motioned to Billy Stone, and he unfolded the chair and put it down near the guy. "Why don't you sit down?" said Cooper. The guy didn't move. Billy reached out and put a hand on his shoulder. Then he eased the guy down into the chair.

There was shouting in the hallway and feet pounding up the stairs and the police burst into the apartment, spreading out quickly to search the other rooms. "Better take that helmet off," said one of the cops. He reached for the helmet and the guy shot to his feet like a rocket, knocking Billy Stone to the floor. We grabbed hold of him and tried to wrestle him back down. His arms were rock-hard. He swung them from side to side, clasping his hands over his head and shouting, "I love you, blue; I love you, blue boy; chromosome, chromosome, chromosome," over and over again. Eventually we wrestled him back down into the chair.

"That's OK," said Davey as he buckled the straps. "The helmet's all right. You can keep the helmet on."

Joe and Billy Stone dressed his arms. The words kept pouring from his mouth: "I love you, I love you, blue boy. Chromosome, chromosome, chromosome." When we carried the chair down the stairs he shook so hard he threw us against the walls.

The crowd closed in for a closer look as we emerged from the building, but we pushed through them. The guy kept repeating himself like a scratched record. Straining in unison, we lifted him up into the Rescue. A short kid, about twenty, stepped forward out of the crowd and said he was the guy's friend. We let him aboard, then pulled out.

We sat on the benches listening to the guy's monotonous chant. Sweat rolled down his face; his blood-smeared shirt was soaked with perspiration. His muscles were taut and his body shaking with the strain of whatever he was feeling. His pulse raced. I leaned forward to check his eyes: they were slightly bloodshot but not at all dilated. He stared straight ahead without focusing. He was in a world of his own, but the vibrations he was throwing off with his rhythmic chant set my own nerves going.

"Do you have any idea what happened to him?" Joe asked his friend. He was a slightly built kid with long hair and thick glasses.

"Are you kiddin', man? He's on a trip. Like this is his tenth or twelfth, I forget. He was in the hospital once before for LSD." The kid grinned at us. "I guess you see a lot of this."

"As a matter of fact," I told him, "this is the first."

With the picture you get from the media, you'd think every other college kid was an acid head or a dope addict. But in a city like Cambridge, where you would expect more problems than in other towns, we just don't get that many drug cases. Of course the Rescue isn't the best measure of the situation. But judging from what I've seen, the drug problem doesn't even begin to compare with that of alcoholism. And although drug cases are on the increase for us, it isn't kids on dope that we're picking up, but adults on tranquilizers. Just a week earlier we'd had an over-tranquilized woman put her car through a store window. They're not supposed to drive on tranquilizers, but they do; and they seem to be able to get hold of them as easily as aspirin.

We took the guy into the Emergency Room. He was still chanting as we transferred him to a bed. Although the kid wasn't resisting, the doctor on duty thought it best to strap him down. Out in the corridor Mary Lane told us that the girl we'd brought in earlier, who had taken the drain cleaner, was in bad shape. They still hadn't identified her, and they didn't expect her to survive the night.

We returned to the house. Around one o'clock we had another run: some drunk's girl friend had locked him out of her apartment and he'd fallen down the stairs and suffered a compound fracture of his upper thigh. We had to put a traction splint on him. The rest of the night was quiet.

Saturday morning I took Larry down to Dunkin Donuts; then we picked up Bev and Helen and did the shopping. It was a warm spring day, and in the afternoon I took the kids to the Stoneham Zoo. We bought bags of animal food and followed the trails up to the deer section so that Larry could feed them and let them lick his hand. Helen wanted to see the seal show but it didn't start for a while, so we wandered over and joined a crowd of people by the elephant house. One of the elephants was being hosed down by her keeper. Her calf had taken cover under her legs, and she looked confused. The kids watched with fascination. I remembered when Larry first came home from the hospital. Bev and I had waited a long time to have a child and then, suddenly, there we were in the house all alone with him. We sat in the kitchen afraid to move for fear of waking him. When you stop to think what it takes for a kid to grow up, you realize what an awesome responsibility it is to bring someone into this world.

At five-thirty Saturday afternoon I was back on the job. The sky was overcast and it looked as if it might rain. We went out almost immediately on a false alarm. Then, around seven o'clock, we had a call to go to the Harvard Square subway, inbound side. "Oh, brother," said Joe as we climbed aboard the Rescue. "Here we go." But Cooper, getting on last, told us it was only an o.d.

Crowds lined the sidewalk under the Harvard Square Cinema's marquee, which announced *Everything You Always Wanted To Know About Sex But . . .* We pulled up at the subway kiosk in the middle of the Square and ran down the stairs to the platform. A priest and a nun were bent over a long-haired, blue-jeaned boy, about eighteen. The boy had been throwing up on the train and he'd passed out when he got off. The priest said he was overdosed on drugs. Cooper examined the kid, then sniffed his breath.

"It's just booze, Father," he said.

We wrapped the boy in a sheet, put him in the chair and brought him up to the Rescue. The kid, still unconscious, sat with his head hung forward, his mouth wide open. His face was red with acne. The instant Dave put the truck in gear the kid started to throw up, heaving vomit over the floor and down his chest. Some of it splattered on Billy Stone's pants, and he quickly wiped it off with a towel. The stink filled the truck.

"It's beer," said Billy.

"No," said Joe, wiping out the boy's mouth, "it's wine."

"No, it's beer," I said.

"Beer with a wine chaser," said Joe with a smile. "Hey Cooper," he yelled up forward. "It's beer."

Cooper looked back over his shoulder. "Michelob or Bud?" he asked.

More of the stuff welled up out of the kid's mouth and ran down onto the sheet. Some of it hung from his chin. Joe cleaned out his mouth, and the kid's eyes jerked open.

"Where the hell am I?" he said. "Hey, I got an appointment. I better make that appointment."

"Yeah, yeah," Joe said. "You're gonna make it."

"Where the fuck're you takin' me?"

"We're just takin' you down to the hospital to get you cleaned up for your appointment."

"Fuck you," he said. "I gotta make that appointment." The kid raised his head and fixed his stare on Billy Stone. "Whassamatter with you?" he asked Billy. "What're ya lookin' at?"

Billy turned away.

"Hey, I'm talkin' to you," said the kid, his head swaying drunkenly from side to side. "You hear me?"

Billy looked at him. The kid drew back his head and spit on Billy. The glob clung to his shirt. Billy wiped it off.

"That's what I think of you, you cocksuckin' bastard," said the kid. "I'll beat the shit out of you, you bastard." The kid's face was red, his fists clenched. We all tried to ignore him. He was wound up tight as a spring. We all knew what he was looking for. He's strapped down and outnumbered three to one, and he savors the situation. If we leave him alone he's shown what a tough kid he is. If we hit him, which is what he'd really like, it does something for him.

We took him into Cambridge Hospital. The staff was irritated and in no mood for games. It was a busy night for them. They'd already had a slashed-wrist suicide, and two stabbings, one in which a guy got stabbed in the belly with a screwdriver by a kid he'd caught trying to steal his car. Almost all the examination beds were occupied and there were other patients outside waiting to be attended to.

We unstrapped the kid. As we lifted him onto the table he began kicking, clawing and screaming at us. We fought to pin him down and nurses and interns came running to help. Mary Lane went for the heavy leather restraining straps, while the others quickly stripped him. He fought like a bastard. There were at least ten of us holding him down. Limb by limb we buckled him to the table.

"Let go of me," he cried, raising his head and shoulders off the bed, his face purple, his eyes bulging with the strain. Suddenly he started pounding his head against the table. A young Puerto Rican orderly grabbed hold of his hair from behind and pulled his head fast.

"Relax. We're here to help you," said the girl with the clipboard. "What's your name?"

"Kirk."

"Kirk what?"

"Kirk."

"Where do you live, Kirk?"

"Let go of me," he shouted, trying to twist his head out of the orderly's grasp. "Tell him to let go of my hair. If he doesn't I won't talk."

The orderly, looking embarrassed, held on to his hair.

"Listen, Kirk," the girl said, "we don't like this any more than you do. It's for your own protection."

The kid twisted his head around to face the girl. "Don't shit me, you fuckin' slut," he said.

The girl turned beet-red. Billy Stone took a step toward the kid but Davey laid a hand on his arm. Two of the nurses ran out into the hall, where a man had just come staggering in with a deep gash over one eye. One of the younger doctors came into the room and told the kid to calm down.

"You tell that Spic to let go of me," said the kid. "You tell him to let me up, Doctor, and I'll talk."

"OK. Let him up," the doctor said. The orderly released his grip on his hair. The kid sat up as far as he could with the restraints binding his arms.

"Let me tell you one thing, Doctor," he snarled. "The fascist pigs think they can put me in chains like a Nigger because I got long hair, but let me tell you, Doctor," he said, cocking back his head, "you're a fuckin' Jew bastard." Then he spat in the doctor's face.

We left the kid alone, except for the orderly, who resumed holding down his head, with the kid shouting obscenities at the top of his lungs. Out in the corridor Joe said to Mary Lane that she should tape his mouth shut.

"I'd love to," she said, "but if he vomited and choked to death I'd get blamed."

Out in the waiting room people were all on their feet trying to see what we could have been doing to the patient to make him scream that much. They looked at us horror-struck as we walked past.

"Another poor misunderstood long-haired youth," said Dave as we climbed aboard the Rescue.

Back at the house Kilroy was trying to get enough guys together to send out for Chinese food.

"Where are you getting it from?" asked Joe.

"I don't know," Kilroy said. "That same place in Boston, I guess."

"That sucks," Joe said. "Every time we get stuff from there I'm sick for days."

"What about that place over in Somerville?" Kilroy said.

"Are you kidding?" said Dave. "That little bastard won't even give us a menu anymore. Every time we ask for a menu he says, 'Solly, no menu Fire 'partment.' "

"Who wants to eat the menu?" said Kilroy. "We can order by heart."

"Then he ups the prices on us," said Dave.

"What's he got against us?" said Jerry Martin. "We only consume half a ton of his stuff a week."

"I don't know," said Dave, "but he means business. The other night I went in there to pick up some stuff on the way home, and as I started to order from memory he handed me a menu. 'Here, Police 'partment,' he says, 'menu.' I says, 'I'm not the Police Department, I'm the Fire Department.' 'Ohhh,' he says, waving his hands. 'No menu Fire 'partment.' And he takes it back. Then, when my order comes, he overcharges me fifty cents. I says to him, 'You overcharged me fifty cents on the fried rice.' He says, 'Flied lice go up Fire 'partment.' I told him I knew what fried rice cost. I picked up the menu off the counter and showed him where it says what fried rice costs. He grabbed the menu and ripped it up and started screaming at me, 'Menu mean nothing Fire 'partment. No menu Fire 'partment!' I was lucky to get out of there alive. I didn't have the nerve to go back in and ask for extra duck sauce."

"How about hamburgers, then?" said Kilroy unhappily.

"Nah," said Joe. "But I might go for some pizza."

"From where?" said Kilroy. "Mario's?"

"That sucks,"

"Anna's, then," said Dave.

"What does a Greek know about making pizza?" said Kilroy. "Right, Larry?"

"It's only the best in town," I said.

"Looks like you're outnumbered," said Dave.

"I may be hungry," said Kilroy, ripping up the sheet on which he'd planned to write down orders, "but I got my pride. I ain't eatin' no Greek pizza."

Just after nine o'clock we had a call: a woman up in North Cambridge said her daughter was very sick.

We got off in front of a single-family dwelling. A woman, about forty-five, answered the door and led us into the back room. The others stayed out in the hallway while Cooper and I went in. The girl looked up apprehensively. She was eighteen or nineteen, with long black hair and dark eyes. I felt her forehead: she was burning up. On the wall above her head there was a painting of Christ.

"Have you been sick long?" Cooper asked her.

"No," she said. "Just today."

"The temperature just started this afternoon," said her mother from the door. "It just kept going up and I got worried. I hope it's all right my calling you."

Cooper said that was what we were there for. We stretchered the girl, covered her with a couple of blankets and brought her out to the Rescue. The mother came with us and sat on the bench across from her daughter. As we pulled out the girl asked me where we were taking her. I told her Mt. Auburn Hospital, and asked her how she felt.

"Not so good," she said faintly.

"I can't understand what's wrong with her," said her mother, dabbing at her eyes with a handkerchief. "This has never happened to a child of mine before."

"Don't worry about it," said Joe. "It's just some kind of bug. She'll be all right."

"I don't know what it is," said the woman. "She just came back today. She went away to see a friend for a week and she just came back and this had to happen."

Joe and Billy and I glanced at each other and—click—the lights went on. I looked down at the girl and she was looking straight into my eyes with a significant little half-smile. I couldn't help but smile back at her. She was a slender, fragile-looking girl, with the smoothest, clearest, marble-white skin. There was an aura of inner beauty and strength about her.

She mumbled something to me, and I leaned over, turning my head so she could talk right into my ear.

"I kinda messed up, didn't I?" she whispered.

"Don't worry about it. It'll work out fine," I said.

"What's that?" said her mother.

"She's gonna be fine," I told her. The mother would find out soon enough for herself. The girl was brave and it was a tough break for her: she'd got pregnant, gone away for an abortion, probably illegal, no doubt paid good money for the damn thing and then got infected as a result of it. I thought of all the girls we'd seen who couldn't have children because as a last resort, in desperation, they'd straightened out coat hangers and performed their own abortions.

The damn thing made your head whirl with the insanity

of it. Abortions should be a matter of choice. It's a girl's whole life, and she should have the option. If the Church wants to advise its followers one way, fine; but, I'm not sure it should attempt to legislate for the entire Commonwealth.

"Poor girl," said Billy Stone as we rode back to the house.

"Yeah," Joe said. "Nice-looking kid, too. Sweet, you know."

"She was beautiful," I said.

"Well, she wasn't exactly no Marilyn Monroe," Finnegan said.

"She had something," said Billy.

"She looked like an angel," I said.

"Come off it, Ferazani," said Finnegan. "Everybody knows angels have blond hair and blue eyes."

"Wrong, Finnegan. Angels have dark hair and brown eyes."

Joe stared at me blankly. "I suppose you like 'em young, too," he said to me after a moment, trying to pick up the conversation. But I was remembering something, and I let the conversation drop.

After my mother died I fell in love with every girl that was nice to me. Usually it lasted only a few hours, but sometimes I was technically going with as many as half a dozen girls at the same time.

My father, brother and I took turns washing and cooking. My brother and I didn't see much of each other. He was always hanging around the track across the river. He dreamed of training horses and living in the country. My father, who had quit his job on the road to become a supervisor in a bakery, was working hard all the time. I respected my father and realized it was hard on him, too, but there was always some strain between us. I felt that I had to learn to be on my own and make my own judgments. The house never felt right without my mother, and I had to struggle to overcome my sense of loss and feelings of insecurity.

I entered Cambridge High and Latin School in 1953. Discipline was less strict than at the nuns' school, and Protestants, we learned, could get into heaven as easily as Catholics. At the time I couldn't have cared less whether any of us made it. I joked around a lot and got in trouble. I went out for the hockey team and didn't make it. I never took home a book, but managed to pass. There was no one to tell me what I had to do, and I did whatever mattered to me.

What mattered was being able to hold up my own end of things. Because of my asthma I was easily winded and couldn't run too far. I hated the feeling of being deficient, so I forced myself to run and swim long distances. I went out for the track team and made it. Not that I lasted too long. They had me practicing for the 50-yard dash, but at the first meet against Arlington our 440 man didn't show up, so they threw me into it. This kid Freddy, he said to me, "Now you gotta go around four times. I'll tell you when to put the speed on." Then the next thing I know, "Get on your mark, get set, bang." I was off. And Jesus, I was running way ahead of the pack. I was going like a sonofabitch. When I came around past the stands on the first lap everybody was talking—they thought I was Flash Gordon or something. When I came around the second time everybody was on their feet cheering. When I came around the third time I ran straight off the track and started puking in front of the stands. I fell over and I thought I'd die. If I had made it around four times I would have set a world's record; instead it was the end of my brief career as a trackman.

I hung out with the guys on my corner. Kerry Corner was 99 percent Irish, and an Italian kid couldn't hang out without being tested. There were always jams and scraps and you couldn't walk away. I bought a punching bag and worked out at the Y for conditioning. I got to be a pretty good fighter, and that mattered to me.

But what mattered to me most of all was having a girl friend. Where we used to hang around depended on where the girls were. If I was chasing or going steady with a certain girl, then I hung around her corner. I hung around a lot of corners. When there was no one in particular we'd go down to the Weeks Bridge and give whoever came by a hard time. We switched girl friends as easily as we abandoned baseballs for footballs in the fall, and footballs for skates in the winter.

One girl changed all that for me. We went together for three years, and during that time she was the most important person in my life. She helped me in lots of ways: little things like how to dress and behave, and bigger things like learning to consider other people's feelings and thinking about what I wanted to do with my life. She had as little experience in life as I had, but somehow she knew much more. She gave me confidence in myself, and she gave me love, which was what I needed most. But we were young, we made a few mistakes, things just didn't work out. Because of her, in every girl I see the beginnings of a woman.

Back at the house we sat around the patrol room watching roller derby from Jersey City. We decided to send out for some pizza after all, and one of the local fire buffs who hang around the house on weekends offered to go pick it up. My mouth started watering in anticipation.

A few minutes later, at quarter to eleven, a North Cambridge box came in. As we were on our way to the location, Fire Alarm reported that the box had been struck for a fire in a vacant building adjacent to a nursing home. We turned off Mass. Ave. down Russell Street and saw the fire coming out of the attic window of a two-and-a-half-story house and lapping up over the sloping mansard roof. We leapt from the Rescue. The Aerial Tower was pulling up behind us; Engine Four was laying a line to the house. Swanson and Finnegan headed around to the rear, while Cooper, Stone and I ran to the front of the house. The door was open. Inside, the rooms were empty.

We climbed to the second floor and turned on our lights to try to see in the dark. The second floor was apparently one large, open room, but the smoke was too thick to see. We felt our way, searching for the stairs to the top floor.

Billy Stone found the stairs, and we started up them. There was no fire showing but the heat was tremendous.

"Holy Jesus, it's hot," said Cooper, halting. "They mustn't have cut the hole in the roof yet."

I was lower than Cooper and Billy Stone but I could really feel the heat on my ears. We heard tramping and voices below us. Engine Four was coming in behind us with the line. They were seized with fits of coughing. We went back down and I grabbed hold of Bob Yanski in the dark; he had the tip and he handed it over to us.

We took the big two-and-a-half-inch line and started to work our way up the stairs again. The Engine Four guys stayed behind us, jockeying line and coughing all the while. The heat grew intense. I could feel it searing the back of my neck; under my coat I was swimming in sweat. But if the fire was above us, the tremendous heat didn't make sense. Even if they didn't have a hole in yet, the heated air banking down shouldn't have been that hot.

Cooper stopped us again. "Something's wrong," he shouted.

"Maybe they're throwing water down through the roof," I shouted back.

"It could be going on the other side of the wall," yelled Billy Stone.

I felt a tug on the line as Cooper moved up a couple of steps. We could see the red now at the top of the stairway. The fire was moving across from left to right. There was plenty of red all around and the heat was murderous. We heard glass crashing; they were hitting the windows. We heard voices as a bunch of guys—possibly the Aerial Company—came in behind Engine Four.

The fire lashed out at us, forcing us down a step. Billy Stone had the tip; I was backing him up. The order was passed to charge the line, and the line grew fat and heavy with water. Cooper gave the order to open up, and I felt the back pressure as the stream of water tore into the red.

From below came a deep rumble, then—BOOM—a thunderous explosion. A searing-hot wind swept past us. The fire swirled crazily at the top of the stairs; then, like a roaring wave, the flames broke all around us.

"Back the hell outta here," screamed Cooper. We retreated down the stairs, dragging the heavy line with us, not knowing what was happening. It was totally black, but there was plenty of red showing overhead. It looked as if part of the attic floor had caved in. There was a good chance that the whole thing would come down. It was too dangerous to be inside.

In the darkness we couldn't find the stairs to the first floor. There should have been a banister but it wasn't there. The fire was raging up above us. I thought how when a ceiling falls the fire falls with it, then suddenly Bob Yanski was shouting that the stairs were gone. "There's just a hole," he yelled.

Someone cried out.

Someone yelled that there was fire below us. Voices were shouting and hollering. I clung to the line. In the darkness I could feel it being tugged in half a dozen different directions.

"Back the line out with you," shouted Cooper above the confusion. "Don't let go of the line." Whatever happened we didn't want to be caught without that line. If we dropped it and ran, the fire might catch up with us. As long as we had a line we could protect our retreat.

Somebody grabbed hold of me. "Who's this?" he yelled through his mask.

"Larry."

"They want you out of here," he shouted—it was Davey. "There's fire below and the whole stairway's gone. They've thrown a ladder for you. Follow me."

We passed the word back and began dragging the heavy line with us, hoping that the fire on the stairway wouldn't burn it away. We all crawled through the darkness to the corner where they'd thrown the ladder. We crawled fast. A fire doesn't start in two places at once unless there's arson involved, and with arson you throw the book out the window, because there's no telling how bad it is, whether the whole place is doused in gasoline or what.

I heard voices moving toward the fire. I recognized Jerry Martin. "Get the hell outta here, Jerry," I shouted. "It's going down under us."

We worked our way over to the window, and one by one, first the truck men, Martin and Mondello, then Engine Four's men, Yanski, Peterson and O'Keefe, got out onto the ladder. They were all coughing, and O'Keefe was choking so bad that Davey had to take him down.

Cooper, Stone and I remained crouching just inside the window. The line had come up over the stairs, which were gone now, and we couldn't jockey it any farther. Billy Stone and Cooper held the tip and opened up on the red. The water shot out into the darkness. It felt good to be behind that line.

The red above us began to darken down. Word came up the ladder that the Deputy said they had a line going now on the first floor. We started moving back in toward the attic stairs. We climbed the stairs carefully because of possible structural damage, got in close to the fire and knocked down a lot of it. They were hitting it from the windows, too, and it began to die. In a few minutes the whole thing was over and the smoke cleared.

While they were doing the overhaul we went out down the ladder to take a look at the site of the lower fire. It had started in a room to the right of the stairway and had been the source of the intense heat we had felt. The walls of the room were blackened and blistered, and where the fire had broken through, the stairway had collapsed into a heap of smoldering ruins. Minutes before, the fire had seriously endangered our lives; now it was just another case for the state Fire Marshal, who investigates all fires of suspicious origin.

There was no question about this fire as far as we were concerned. We knew it was arson. We didn't want to think about it now, though, not so soon after we had almost been beaten by the flames. It's almost impossible to convict an arsonist unless you catch him in the act, and that rarely happens.

The Deputy released the Rescue. Outside we were surprised to find that it was raining. The rain felt wonderfully cool and refreshing after the heat and smoke of the fire. We stood for a minute with our helmets and coats off while in the truck Joe finished treating one of the engine men for a burn on the back of his neck. Then we returned to the house, washed the soot off our faces and went upstairs to the kitchen, where boxes of cold pizza were waiting for us.

Dave wanted to reheat the pizza but the rest of us were

too hungry to wait. "Jeez," he said, "I don't know what's wrong with you guys. You eat your pizza cold, and you're always getting into jams. I'm getting tired of saving you guys."

We laughed. We had been damn glad to see him, but any one of us would have done the same thing. We no longer counted the number of times we'd pulled each other out of tight spots. We always looked out for each other in a fire, and none of us ever left a burning building without first making sure that the others were safe. There wasn't one of them I wouldn't have trusted with my life.

The skin on the back of my neck felt tight; looking at Cooper and Billy Stone, I saw that their necks and ears were red from the heat we'd encountered on the stairs.

"I forgot to tell you," said Cooper. "The Deputy said to thank you for a good job."

"You're very welcome," said Dave.

"I got the impression he meant all of us," Cooper said.

"Yeah, well," kidded Dave, "I'm thinking of asking him for a raise, what with all his people I've been saving around here."

We joked around about what had happened, feeling close in that tired, satisfied way that you do after a fire, flushed with the feeling of respect and affection you have for people you've been through something with, especially when you've relied heavily on each other. And there was also the satisfaction and self-respect that come from knowing that others have counted on you and you have held up your end. That was particularly important to me. Unlike many Cambridge firemen, who come from families with fire-fighting experience and who know about the job and how dangerous it is, I had come from the outside and had to earn my acceptance.

It was after one o'clock. Cooper stood up and yawned.

Outside we could hear it raining hard. "The rain'll bring 'em out," said Finnegan. "They'll all be gettin' out of the bars now." Cooper said he wasn't waiting around for it to happen. He and Billy Stone sacked out. Dave, Finnegan and I sat up past one-thirty, then we went to bed. I was exhausted; I thought I could sleep forever.

The phone rang. The lights flashed on. I squinted at the wall clock as I pulled on my clothes; it was four A.M. We hit the poles. "Report of a bad accident," shouted Cooper as we ran for the Rescue.

We sped up Cambridge Street. The wind rushed in through the open rear doors and the road shot out from under us. The rain fell steadily onto the slick black streets. The city looked deserted.

We turned down Clark Ave., the grim street where all the city's auto junkyards are. Ahead of us we saw the flashing blue lights of the police cars, then the wreck itself.

"Holy Jesus," Joe said. "Look at the telephone pole."

The car had smashed head on into the pole, and the top part of the pole hung like a snapped twig.

We pulled to a stop. As we ran to the car I could see a kid inside, right up against the back window. One of the cops said it was a stolen car. "There's two of 'em here," he said. "The others took off."

I could see the other kid slumped forward on the passenger side of the front seat. Cooper and Finnegan went for him; I took the driver's side.

The whole front end was bent around the pole. There was glass scattered everywhere from the broken windshield, and the side panels were buckled. The driver's door was open and the seat thrown forward. I squeezed into the back to get at the kid. He was a young boy, about fifteen, and he was lying faceup in the compartment behind the rear seat, his nose pressed against the window. I didn't see how he had gotten back there. The impact should have thrown him forward. He looked perfectly natural. There wasn't a mark on him. I examined his eyes and felt for the pulse at his neck, but I knew before I touched him that he was dead.

The kid in the front seat was crying to himself. Cooper and Finnegan were working on him. He'd gone into the windshield and he was cut up pretty badly. Blood was running from his mouth. He had multiple facial wounds, a couple of broken teeth and a fractured arm.

Cooper looked up at me questioningly. I shook my head.

"OK," he said. "Get the jiffy splint and the collar."

I got the equipment for Cooper and stood by to assist. Behind the fences on either side of the street rose dark heaps of scrap metal and wrecked automobiles. Through the rain I watched a cop marching two boys down the middle of the street. They were both covered with glass and their faces streamed blood. One of them was limping badly and holding his arm.

Billy and I went up to them. They couldn't have been more than fourteen or fifteen. "Here, you take 'em," the cop said, letting go of their arms. "I hope they're proud of themselves."

"You know your friend's dead," the other cop said.

The kids said nothing, just stood there with their heads lowered.

"Whatsamatter?" said the first cop. "Now you've got nothing to say?"

"OK," I said. "Why don't you let us take them now."

We brought them over to the truck and told them to sit down inside. Their faces were messed up, but they didn't have any broken bones or teeth. The one I examined had a deep gash on one side of his head. I checked him over for a possible concussion. It's hard to judge the condition of people after an accident; the fact that they were walking meant nothing.

"Where does it hurt?" I asked the boy, examining the

rest of his body. He was very young and didn't look to me like a bad kid.

"It doesn't," he said with a shrug.

"How old are you?"

"Thirteen."

"And your friend?"

He didn't answer. The other kid was silent, too, while Billy Stone splinted his arm. They weren't interested in their injuries. Most people at accidents want to talk. They worry about their families, their car, their wallets. They want medical treatment and sympathy. They want to talk and to be reassured. But these kids had nothing to say.

They both looked up as Cooper and Dave slid the stretcher with the sheet-covered body onto the Rescue. Then Finnegan rolled up the chair with the kid who'd been in the front seat, and we all lifted him aboard and pulled out for Cambridge Hospital.

Inside the Rescue no one spoke. We sped through the heart of the city, the streets wet, cold and black. In Inman Square we swept past a newspaper truck unloading the Sunday *Globe*. A few hours later people would be reading all those papers filled with all those words. At four o'clock in the morning there may have been words, but we had no will to speak them.

=*SEVEN*=

I WAS UP IN HARVARD SQUARE DOING SOME shopping on a hot June afternoon, when I saw this guy coming down the street past the Coop. He had a little swagger, as though he were half in the bag, but there was something else about him, something familiar. I stopped. It was Ken O'Donald.

"Hey, Ducky!" I shouted.

He looked up and did a double take. "Hey," he cried, grabbing my hand. "How the hell are you?"

We stood there grinning at each other. It had been over ten years. We hadn't seen each other since high school.

"Jeez," he said. "You haven't changed a bit."

"You neither. Where are you living?"

"Over in Charlestown," he said.

"Married? Kids?"

"Yeah," he said. "Three."

"Who'd you marry?"

"You don't know her," he said. "She was from Charlestown."

"I heard you were selling."

"Yeah," he said, looking away. "I'm making out all right. How about you? Made a million yet?"

"No. I fell short a few hundred Gs," I said, thinking that if anyone had figured to make a million it was Ducky. He had had all the big plans. He was going to go to school and become an architect or an engineer. He was going to strike it rich. As a kid way back in grammar school he was always full of moneymaking schemes that involved going up to Brattle Street to mow lawns, rake leaves or shovel snow. But the people up there had their own gardeners and maintenance men and didn't want us hanging around. Every time we went up there hoping to make a few bucks we came back with beans.

But Ducky never got discouraged. And he was always building things like tree houses or snow forts. I remembered that Ducky had known how to pack a snowball around a core of ice so that it really stung, and that he had had a steady arm for throwing them.

Now his hands were shaking just a little, and he held one arm stiffly.

"Looks like you've got a few under your belt," I said. "Where the hell have you been drinking?"

"Up at the O.G.," he said. "I come down to see my mother."

"Hey, does Terry Moran still live up above your mother's?"

"Yeah, I guess so."

"What about the rest of the gang? Have you seen any of the other guys?"

"I'm livin' over in Charlestown," he said, rubbing his arm. "I don't see nobody anymore." He was perspiring in the midday heat, and he wiped the back of his sleeve across his forehead. Then he put his hands in his pockets and looked away.

"What's the matter, Ken?" I said.

"Aw, Jesus," he said, lowering his head, "don't even talk about it. I'm all screwed up. My old lady just threw me out this morning. Slammed the goddamn door on my arm. Hurts like a bastard," he said, rubbing his elbow.

I told him what the hell, she'd cool off.

"Nah. She's already thrown me out a couple of times. She's right, too. That's the worst part of it." He put a hand on his back pocket. "Look," he said suddenly, "I'm on my way back up there now but I haven't got any money with me."

I gave him a couple of bucks and we crossed over to the taxi stand by the kiosk. I felt sorry. He'd been a carefree guy and it made me sad to see what the years had done to him.

"I'll give this back to you," he said, waving the bills at me as he climbed into the cab.

"Forget it," I said. "Good luck to you." I hadn't seen him in ten years and I didn't expect ever to see him again.

I was wrong.

A couple of days later I was working the Thursday night shift. We'd had a few minor incidents. An entire family was trapped in an elevator when the superintendent cut the power—he said they were five months behind on the rent and trying to skip out. Then an elderly woman called us down because her cat wasn't feeling well, a drunken woman collapsed in a phone booth, and we had an epileptic in grand mal seizure whose tongue was cut up pretty badly, only because someone had put a spoon over his tongue instead of between his teeth. They'd done more damage than if they had left him alone.

Around ten o'clock we had a call to go to an address in back of Harvard Square. I recognized the three-story house as soon as we pulled up in front of it. It was Ken O'Donald's mother's place. Terry Moran, another old friend, who had always lived up over the O'Donalds, was waiting at the door.

"Larry!" he exclaimed, surprised to see me. "It's Ducky. He's swallowed some sleeping pills."

We followed Terry through the living room, where Mrs. O'Donald sat crying, into Ken's old bedroom. Ken was lying semiconscious on the bed, moaning softly. The bottle of pills was beside him on the night table. Cooper and Dave examined him: his eyes, pulse and respiration were all good. We put him in the chair and rushed him out to the Rescue. Terry came along with us.

On the truck going down to the hospital Dave and Joe kept shaking Ken and talking to him, trying to keep him awake. Apparently Ken had spent the last couple of days sleeping on Terry's living room couch. He'd been despondent, Terry said. Things had gone badly for him in business; he'd missed out on a couple of promotions, gotten into debt and started hitting the bottle. His wife had finally kicked him out. It was a long story, Terry said.

Ken opened his eyes halfway and started mumbling. "Hey, Ducky," said Terry, turning Ken's face in my direction. "You know who this is? Huh? Look who's here."

Ken looked up foggily and nodded. He mumbled something that could have been my name. Then he closed his eyes.

He was still semiconscious when we brought him in. They hooked up the suction machine and started pumping him out. He'd live to face another day.

A little while later, out by the desk in the corridor, Joe asked me what had happened to him. I told him I thought it was money troubles.

"That's what's wrong with this country," Dave said. "Everything stems from the buck. People are killing themselves trying to make the buck. Everybody knows it, but they keep on doing it just the same."

An orderly rolled out Ken O'Donald's bed and pushed him slowly down the corridor and around the corner, out of sight.

When I knew him, Ken O'Donald was always full of life, full of plans. He was going to strike it rich.

And so was I. When I graduated from high school I wanted to make money right away. I figured sales was the best way to do it without going through the college rat race. I thought about college; I was even accepted at one place over in Boston. But I was anxious to make money and be independent. I figured I could always go later. I decided I'd work awhile and put some money aside. I figured I'd blaze my own trail to success.

My first jobs weren't much. I worked in the claims department of a large insurance company. The job paid poorly and offered no room for advancement. Next I worked in a men's clothing store up in Central Square. It was just a living. I figured these jobs were only stepping-stones.

The best decision I made during this period was to apply for a two-year Officer's Candidate program: I took a written exam, went before an Army board and was accepted. In my second year the Berlin crisis broke, and because they expected that the Yankee Division would be activated, they accelerated our program. They stationed us full time at the Commonwealth Armory, where they brought in special instructors who crammed us with material. I studied hard, graduated high in my class and received my commission; but in the end we weren't activated.

When I returned to civilian life I landed a job as a life insurance salesman. Bev and I had met and were already engaged, but before we got married I wanted to establish myself in a position where I could make some money. I figured this job was it.

The company gave me what they call a debit—that's a certain number of established accounts. My job was to maintain those accounts and increase the insurance in my territory. I got seventy-five bucks weekly as a base salary for the collection of premiums on existing accounts, plus a conservation bonus simply for the continuation of those accounts. Over one hundred bucks a week before I even started selling!

Then, for every dollar's worth of insurance I sold, I was to get a dollar for the first twenty-six weeks, and a smaller percentage over the next twenty-six weeks. To make real money you had to keep selling, but I knew I could sell anything to anyone. Especially something, like life insurance, that I genuinely believed in. I figured I was on easy street.

The debit they gave me was poor: the all-black community down through Western Ave. and Howard Street. They told me not to worry. There was a good buck to be made there, they said.

Early every morning I went down to my territory. There was a lot of life on the streets and the people were warm and friendly. They were no different from the poor people I had known growing up on Kerry Corner. It was good business to visit with my clients, and I genuinely enjoyed having coffee with them at their kitchen tables, shooting the breeze and listening to their troubles. Most of them had weekly insurance: they paid premiums anywhere from a nickel up. If they missed four weeks in a row they lapsed.

If they lapsed you had to reinstate them, which was bad for them because it meant in effect that they were taking out a loan against their policy, and if they made a claim during that period they had less coming than they thought.

The people in my debit were poor; often they couldn't even pay for their groceries. I started lending them a buck here and there out of my own pocket. After a while I was lending money left and right.

But the company wasn't paying you to be Mr. Nice Guy.

They wanted you to increase your client's insurance. They didn't tell you you had to ram the policy down the guy's throat. They just set things up so that for you to make any money, you had to be self-motivated.

It was easy enough to do. There you are sitting with the father at the kitchen table. His job isn't too good and he has three or four kids and another on the way. It isn't hard to convince him he could use more insurance. If he needs convincing you just push his face up against the ambulance window and show him what can happen. The guy's already managing to pay a buck a week, so you figure he can probably scrape up another buck or two.

You show him everything he'll get for that extra buck. You make it simple for him by skipping all that fine print. You hit him with a lot of numbers but keep quoting the highest figure. You hold out prospects of the gold mine. You play on his fears, his greed, his love and his deep underlying sense of having failed to do as much for his family as he wanted to. It's easy enough to do.

I couldn't do it. I could sell all right, and I sold a lot of good policies to people who had the money. But I didn't have the heart to oversell insurance to people who couldn't afford it. You had to be a hard-nosed bastard to sit down with poor people at their kitchen table, drink their coffee and then force something down their throats because it meant an extra buck in your pocket. But there were plenty of guys who could do it just fine.

I remember one young couple in particular. They had just gotten married and the wife was pregnant. She had big shining eyes and every time he threw an arm around her waist she laughed out loud. I liked them a lot. I sat down with him and we talked about his prospects. He had a small income but big ideas: he wanted to insure himself, his wife and his child when it was born.

I leveled with him. I told him it wasn't worth it at that point for him to insure all of them. He could always take care of himself; his earning power would continue. But if anything happened to him, their source of support would be lost. The important thing was to protect them by insuring himself. I told him a buck a week was all he really needed to spend. Then I spent two hours with him explaining the policy. As soon as I returned to the office to write it up I got a call that he wanted to see me right away.

I went back to his place.

"You can cancel that policy," he said. "I just bought me a better policy."

"What policy?" I said. "Who'd you buy it off of?"

He named a guy who worked for a rival company. He was one of those who did very well for themselves. He had shown up the minute I left.

"Just look here," he said, spreading out the company description of the policy on the table. "It's only four dollars a week and look at all the money it gives me."

With a sinking feeling I sat down at the table to read the fine print. He had bought the gold mine—ten thousand dollars' worth. He couldn't afford it in the first place, didn't need it in the second and wasn't going to get it in the last. It was one of those three-thousand-dollar base policies that pay triple indemnity if you get shot on a one-way street in front of a three-family house overlooking a river.

"You *see*," he said, slapping his side and grinning first at me, then at his wife. "Ten thousand. Written in black and white."

"Jesus, look, this is only a three-thousand-dollar base policy," I said. "If you die your wife gets three thousand."

"No she don't," he said, slamming his fist down on the table. "Don't you shit me. She gets ten."

I got up from the table and went to the door. "She gets

ten if you get shot on a one-way street," I told him. "Good luck to you."

I was too mad to go back to the office. I was mad at myself for trying to help someone who wanted the gold mine, disgusted with the salesman who'd sold it to him, fed up with the cycle of poverty that trapped people and the system that allowed others to prey on them. A few days later I went in and quit. Once again I had no job and no prospects.

We'd picked up Ken O'Donald on a Thursday night. The following night I was detailed out to Engine Four on Mass. Ave. up above Porter Square. Engine Four had two guys out on injured leave, and they were shorthanded.

It was nice to spend some time in another house talking with friends I didn't get a chance to see much of. It was certainly a lot quieter than being on the Rescue. Engine Companies in Cambridge average only a fourth as many runs as the Rescue. Which is not to say that theirs is an easier job. Their work lies elsewhere. I knew because I'd been on Engine Three down in East Cambridge before getting on the Rescue.

The first Engine Company into a fire takes a real beating. The first ten minutes—laying the lines, jockeying them into position and working inside without the masks—are an ordeal. It's as intense and strenuous as a boxing match, only the tension is worse because the risks are greater and you haven't had time to prepare yourself. If you're the first alarm you have to stay through the duration of the fire, and after the fire you make up lines and assist in the overhaul of the building.

When you return to the house you have to drain and hang the wet hose in the towers to dry, then pack dry hose in the wagon bed. After the exertion of working a fire, those hoses feel as heavy as iron pipes, but you have to get your Company back in service as fast as possible. Then there's the daily routine of housework, drills and prefire planning. There are also the special seasonal chores. In the winter after snowfalls you have to shovel out every single hydrant in your district. In the summer you inspect the hydrants, as well as half a thousand homes and businesses in your area. There's plenty of work on an Engine Company, only the pace is different from the Rescue's.

It was a muggy summer evening. The crickets were chirping away and occasionally a firecracker would explode. A few neighborhood kids were riding their bikes on the driveway out back where Pat Peterson was working on his car. We stood around watching him, joking and drinking tonics. John Murphy came out to get up a list for southern fried chicken. John was a stocky, potbellied old-timer, always complaining that things weren't as good as in the old days. He wrote down the order slowly, leaning carefully over the paper and holding the pencil like a spear.

"John never learned to write down orders too good," Bob Yanski said. "When he first come to this house all of North Cambridge was woods, and for supper they'd just go out and shoot squirrels."

"Dammit," said John. "You made me break the point. Can you kindly wait till I'm finished takin' the order before you're startin' up with me."

Bob Yanski and I walked around to the front of the house. The sky was streaked with pink and orange and split by occasional flashes of heat lightning. We stood in the fading twilight, smoking cigarettes and watching the traffic roll by. Directly across Mass. Ave., a few doors down from Gator's Bar, there was a guy in a leather jacket leaning on the mailbox.

"He's out there all year long," said Bob. "The other day the kids that live up in the building lowered a can of beer down to him. He drank the beer and threw the can in the mailbox. Throws all his garbage in there. If he's gonna spit, he spits right inside the mailbox. Does everything right there. He's out there all year. In the winter he dresses exactly the same as he does now. He knows the whole crowd that goes into Gator's."

Gator's draws a rough crowd. There was an underworld slaying there a few years back: two body guards were shot in their parked car out in front. The guys at Engine Four are experts on that killing the way some people know about the Civil War, or the way Joe knows about the Amazon. Dave is always saying that Engine Four should put together a tour that includes a talk and two free drinks at Gator's. You can't spend any time at Engine Four without at least one of the men getting around to the shooting.

"Was your group in on the shootin', Larry?" Bob asked.

"No, Group Four caught that one."

"He got 'em with a high-powered rifle," said Bob. He blew a stream of smoke through his nostrils and gritted his teeth in gangster fashion. "You guys down there at Harvard don't know what life's all about," he said.

Night fell. Things are slower in the summer. You get grass fires and car fires and rubbish fires, but you don't get as many building fires as you do in the colder months, when heating systems are an additional hazard. Which is just as well; nothing exhausts you faster than fighting a fire in severe heat and humidity.

We had two runs all evening, a rubbish fire up at the Jefferson Park Project around nine-thirty, then a false alarm a few minutes later at the same place. I was hoping to see the Rescue, but they got the "All Out" over the radio en route. We got off to check out the false alarm in front of a little pocket park. The heavy night air smelled of hot asphalt. There was a pickup basketball game going on by the light of the streetlamps, and a row of kids sitting on a stone wall with their backs to us, their bodies deliberately arranged in postures of watching the game. Fire trucks with sirens and flashing lights were moving up the street right behind them, but not a single head turned. Usually they were chasing us down the streets. What could you do?

At quarter to eleven we sat down with a bucket of chicken from Colonel Sanders. Old John Murphy started shoveling mashed potatoes into his mouth as if he were stoking a fire.

"Hey, John," said Yanski. "When you get to the red part, stop. That's the plate."

"You can eat the box, though," said Peterson. "That's biodegradable."

John looked up from his plate and gave them both the finger.

"Poor John," said Yanski to me. "You hafta excuse his manners. He's not used to flat plates. In his home they have separate compartments to keep you from mixing your food."

John muttered to himself and reached for another chicken breast.

"Larry, this is a happy house," Yanski said. "We got a good house here."

After supper we drank more tonics and caught the late news, the ball scores, the weekend weather forecast and the start of the late movie. Then they all headed up to bed, except for Yanski, who had the floor patrol.

I wasn't tired. I sat on the back step of the pumper and wondered what kind of night the Rescue was having. The pumper and the wagon, their cab doors thrown open, were side by side, like tame animals in a stable. It still struck me as incredible that my life was bounded and shaped by these unique red trucks. When I first came on the Department I wasn't even sure of the difference between a ladder and an engine truck. They both seemed to me like fantastic glittering toys. I had to take some guy aside and ask him; I thought his eyes were going to pop out of his head. I didn't know one practical thing about the fire business. All I'd done was study the Red Book, the standard fireman's text, and take the exam. Before that I had never given the Fire Department a second thought.

It was my grandfather who first suggested it. I had just quit the life insurance company and Bev and I had only recently been married. We had moved in next door to my grandfather in the duplex on Harvard Street and I talked with him a lot about my job situation.

"Why not join the Fire Department?" he said one day completely out of the blue.

"The *Fire Department?*"

"Why not? It's a job with security. And you'd be doing something worthwhile, too."

"I couldn't be a fireman."

"Why not? You could take the test at least. It's a very secure position."

My grandfather had been with the civil service for over twenty years as a supervisor of mechanics for the Department of Public Works. Job security was very important to him. I told him it wasn't my primary concern. I just didn't want to have to screw anyone to make a decent buck. I put the idea completely out of my mind.

Instead I took a job as a salesman for a major food company. I didn't have to face the helplessness of poverty every day, and I had a chance to earn good money. My job was merchandising food products to warehouse and chain-store accounts. In the chain stores—such as A & P, First National, Stop & Shop—the object was to enlarge the amount of shelf space given to our products. If there was a new line coming out, or a new item in our old line, say, cherry-apple juice with gumdrops in it, then I'd try to sell it to the grocery manager. The key to it was making friends with the manager.

Competition among rival food companies was keen, but I filled my quotas easily and won special bonuses. After four years on the job I was in line for promotion to district supervisor. I was finally on my way.

There were drawbacks, of course. There are a lot of petty irritations in a salesman's life. The store manager is your bread and butter, and you have to put on a chorus-girl smile when you go in to see him. A lot of days you don't feel like smiling, but you have to build up a rapport with the guy. The key to selling is selling yourself. If you can't sell yourself, your product is dead. That's easy enough to remember when you like the guy, but a lot of them aren't at all likable. Some of them take advantage of their position and try to shake you down for a few extra cases because of "breakage," or they tell you their friend's having a baby and why don't you take care of him. You'd like to send them six feet below but you have to give them the chorus-girl smile.

The job comes home with you, too. There are reports to complete and mail. There are calls at ten P.M. to discuss your day's sales. Then there's the supervisor who gets up at six in the morning. He thinks everybody gets up at six in the morning, so he calls you at six. The phone rings and you stumble out of bed thinking it's some big emergency. "Look," says a shrill voice at the other end, "Where's the banana-apple-cherry juice?"

"What banana-apple-cherry juice?"

"Last night I was out at the Stop & Shop. They don't have any banana-apple-cherry juice."

You tell the guy you've ordered it. You tell him you can't understand what's happened to it. You tell him you're very upset about it. Yessir, yessir. You tell him you'll look into it right away.

It's the same charade day after day, and after a while, in spite of the money, it gets to be a real drag.

As the time for my promotion drew near I couldn't help but wonder where it was all getting me. Except for money, the job gave me nothing. At the end of every day I felt hollow. I began to dread each morning; I felt tired and rundown, and unable to look people in the eye. The thought that I would have to go on doing the same thing made me sick. Where was I going? What would I have accomplished in the end if I sold fifty million cases of baby food and they gave me a gold watch? Maybe I would drive a Cadillac, live in a hundred-thousand-dollar house, force my kids to take tennis lessons? It had never struck me as necessary to be rich in order to enjoy life. I just wanted to earn enough to have what I needed without worrying that the bill collectors would come pounding at my door.

I didn't think I could face the prospect of a lifetime of work that was of no real worth to anyone and that gave me no sense of personal fulfillment. I started casting about. I read *Death of a Salesman* and that made it worse. I talked to a few friends about what I felt. They admitted they had the same concerns, but they advised me not to throw away a good promotion. There wasn't anything you could do about it, they said. It was just part of the scheme of things.

I returned to the Rescue on Monday morning. It was good to see the guys after the weekend. Davey had been up to New Hampshire to work on his house. Joe had worked on his car. Billy Stone had finally gotten around to painting his rabbit hutch. Cooper was going around smirking like the cat that swallowed the canary. Dave said you could see the feathers sticking out of his mouth. The new promotion list had just come out and Billy was next in line for Captain.

We held drills, did our housework and completed an

inventory of the equipment aboard the Rescue. It was a hot, sticky, oppressive morning and nobody was moving too fast. We had one run all morning, down to an office building off Central Square. The office supervisor met us at the front door and led us inside past rows of desks. The people hardly looked up from their work.

In a hot, dusty back room crammed with four desks, a man sat slumped forward mumbling to himself. He was about twenty-seven, tall, thin and nervous-looking. In front of him on the desk were papers, an adding machine and an incoming-outgoing box. Overhead a large-bladed, old-fashioned fan whirred. One of the men said that the guy had been all right when he came in that morning; then, while he was working on his adding machine, he'd just slumped forward and started mumbling. He'd been that way ever since.

We walked slowly up to the man. He was perspiring profusely and responded incoherently to Cooper's questions. Billy told him that we were going to put him in the chair and strap him in just so he wouldn't fall out, and he offered no resistance. As we wheeled him out through the main office, the typewriters were clattering briskly all around us.

On the way to the hospital the man began to mumble louder and faster. He jerked his head from side to side like the carriage of a typewriter knocking off a hundred words a minute. His body was taut and beads of sweat rolled down his cheeks. He had taken hold of one of the straps that bound his legs and while he mumbled he ran his fingers up and down it as though he were adding columns of figures.

I asked Joe what the guy's first name was. He shrugged his shoulders and called up forward to Cooper. It was Robert.

"Hey, Robert," I said. "Hey, Bob."

The guy looked up at me, still mumbling, still jerking his head from side to side, still running his fingers over the strap.

"Hey, Bob, take a break," I said. He tilted his head to one side and looked puzzled. "It's time for your coffee break," I said.

He shut his eyes, nodded once and relaxed. We didn't have another sound out of him all the way to the hospital. In the Emergency Room they called for a psychiatrist.

On the way back to the house Dave said, "I suppose you think you're some kind of shrink, now."

"That's right," I said. "The only one in town with a portable couch for house calls."

"Well, I suppose I can't afford not to try you," Dave said, lying down on the stretcher. "Should he be here?" he asked, gesturing toward Finnegan.

"He's OK," I said. "That's my secretary."

"He's awful sweet-looking," said Dave with a knowing wink.

"Aw, Jesus," said Joe, turning his back on us.

"Proceed," I said.

"Well, it's this man where I work," said Dave. "From the moment I laid eyes on him I haven't been able to think of anything else, but he won't give me the time of day. He's not much on looks; he resembles a cross between Fu Manchu and Oliver Hardy."

"Aw, Christ," said Joe, moving up forward with Cooper and Stone.

For lunch we sent out for submarines. We took them upstairs to the kitchen and got ice-cold tonics from the machine. As we were unwrapping the sandwiches and taking our first satisfying bites, the loudspeaker sounded. *"Attention. Rescue going out."*

The call was for a heat-exhaustion victim at a supermarket in Central Square. The store manager waved nervously in the general direction of the meat counter, where a crowd of shoppers had gathered. In the center of them, lying facedown, was an enormous fat man, about fifty.

"Oh Jesus," Dave said. "Fat Harold's pulled his stunt again."

Cooper and Joe knelt down beside him. "Get up, you son of a bitch," said Cooper softly. "It's too hot today." Joe leaned over Harold's ear and whispered something. Harold started to groan very loudly.

"Why don't you do something for that poor man?" said one woman.

"Can't you see he's suffering from the heat?" said another.

Harold groaned. We all looked at each other helplessly. Finally, realizing that Harold had us over a barrel, Dave went out for a stretcher.

Harold weighs three hundred pounds, give or take a dozen. He lives alone behind the Cambridge Hospital, and on nice days he goes window-shopping, then hitchhikes home. You see him hitchhiking all over town. But some days his luck isn't good, and especially in hot weather Harold hates to wait. So he goes to the nearest crowded public place—a supermarket, a dime store, occasionally a discount jewelry shop—and falls down and starts groaning. Somebody calls the police and the police call us. Since there's always a crowd we have to give him service.

Dave and Billy brought the stretcher. It took all five of us plus two cops to dump Harold onto it. As we hauled him through the check-out counter the girl behind the cash register gaped.

"Don't be alarmed," Dave told her. "We're just throwing a banquet for a few hundred cannibals." We laughed so hard we nearly dropped him. By the time we got him loaded into the truck we were sopping with sweat.

"God bless youse boys," said Harold as we pulled out. "Bless you, ah bless you."

"Harold," Dave said, still panting from carrying him,

"anybody your size should be made with handles, for easy lifting."

"Oh bless youse boys," cried Harold. "Bless you, bless you." He kept it up all the way to the hospital. Not that it did any good to take him down there. Harold is a Christian Scientist and won't accept medical treatment.

"Harold," said Dave as we pulled into the hospital driveway, "would you like to get off here, or would you prefer that we carry you inside first?"

"Bless you," said Harold. That's all he ever says to us.

We lugged him into the accident room and dropped him on a bed.

"Oh no, not again," said Mary Lane. Without another word to anybody Harold got up and waddled out across the parking lot to his home.

We started back to the house. We get a lot of junk calls involving fakers like Harold. There was one local alcoholic who used to go on periodic binges. He'd check into a hotel, drink the bar dry, then fake a heart attack. There's no way of telling at the scene whether or not someone is faking something like that, and we'd have to take him down. People in bars fake heart attacks all the time. In the city jails, too. It's annoying, but it's often sad; it reminds you of all the lonely, troubled people who just want someone, anyone, to reach out to them.

I remember the first time I experienced the feeling of being the one to reach out. I was in upstate New York for two weeks of active duty in the Reserves. I had just transferred into the Medical Service Corps after training with the Army Medical Field School, because I had gotten as tired of being an infantry line officer and running over hills as I had been of selling baby food.

As a Medical Corps officer I was appointed registrar for our hospital unit, an administrative job that entailed screening and processing medical records and evacuating patients to other military posts.

One of my first evacuees was an Indian who had voluntarily surrendered himself for heroin addiction and was being transferred to a detoxification center.

I remember standing in a stiff wind on the flight line with him and another kid, waiting for the plane to land. The Indian's uniform was starched stiff and he had a shirt full of ribbons. His high cheekbones, his jet-black hair and the proud way he stood gazing across the field made me think of chiefs and warriors. When we were kids growing up on Kerry Corner we used to play cowboys and Indians. It was best to be a cowboy, because the Indians always got shot; but if you had to be an Indian, at least you could always get up and be a cowboy in the next game. There was no such easy out for this Indian.

"For how long," he asked me, "will I be gone?"

I told him I didn't know.

The other kid said, "Gee, I've never been in a plane before."

"I have," said the Indian. "And I have jumped out of them. It's nothing."

The plane landed, taxied up and dropped a stairway. I shook hands with the Indian and wished him good luck. The plane took off and I watched until it disappeared. I wished there was something more I could do for him.

At camp I had a lot of time to think about where my life was heading. I wasn't proud of what I was doing. Thinking of the business world from that distance, I realized I didn't want to go back to it. I didn't know what I wanted to do. I only knew that I wanted to be proud of the way I spent my life.

When I became a civilian again that fall I started looking into other fields. I figured I would line up something first, before I broke the news to Bev. As one option I resolved to take the civil service exam for firemen. I went to my grandfather and he told me to get hold of the Red Book and study it hard.

One night as Bev and I were sitting down to supper I told her I was going to take the fire exam.

"The fire exam?" she said, reaching for her water glass. "What are you talking about?"

"You know," I said. "The civil service exam you take to be a fireman."

She burst out laughing. "That's ridiculous," she said. "You're going to be a supervisor. You can't be a fireman."

"Why not?"

"Well, for one thing," she said, still laughing, "you don't even look like a fireman."

"And what are they supposed to look like?"

"Mr. Foley's brother is a fireman," she said, spreading butter on a slice of bread. "He says all they do is sit around and talk about the price of meat."

"You think those guys don't do any work?"

"I suppose they might get a fire every now and then," she said with a shrug. Then for the first time she took a good look at me and put down her fork and knife. "You're really serious, aren't you?"

"It's a job," I said.

"But you've got a job. A good job." I could see she was getting worried. She knew how headstrong I could be.

"Look," I said, "it's just something I'm thinking about. All I'm going to do is take the exam."

"But why? The job you've got is a perfectly good one."

"Because I'm fed up with it, that's why. I want to do something that'll give me a little satisfaction."

"OK, OK, I know you're fed up. But couldn't you at

least wait till you're a supervisor to decide? Why throw that away? It might be different."

"No, it'll be the same thing," I said. "Only they'll pay me more so it'll be harder to quit."

"OK, listen," she said, pushing away her plate. "Maybe you do need a change. But why the Fire Department? Do you really know anything about it?"

"No."

"Well then, what makes you think that it'll be any more satisfying?"

She had me there. I only knew that I had to try something different, that I couldn't continue my life along the old lines.

I studied hard and passed the written exam. I ran, lifted weights and climbed ropes at the Y in preparation for the physical exam. After I passed that I waited for the city to notify me of an opening. The whole process gave me a new lease on life. The more I thought about learning to be a fire fighter, the more I liked the idea.

When I went in to give notice to my boss at the food company, he smiled, nodded and asked me what my present salary was. I told him $7,000 plus commissions. He offered me a raise on the spot. I told him it wasn't the money. He reminded me that in a short time my promotion would be coming through. He asked me how much I could possibly make as a fireman. Now the starting salary's almost $9,800; then it was $5,800.

"What the hell's the matter with you?" he said. "You'll be making at least twice that much as a supervisor." He said he just didn't understand. He told me I was crazy. But he added that when I was ready to come back my old job would still be there. I thanked him, but told him that in all truth it wasn't very likely I'd be back.

"We'll see," he said. Then he looked at me and shook his

head. "A fireman," he said. "Jee-sus. Not that I have anything against them. I just don't think you're cut out to be one."

The next month, when I reported for duty and found out what fire fighting was all about, it occurred to me more than once that he might have been right.

We returned to the house and finished our partly eaten sandwiches. During the afternoon there was a box alarm for a small grass fire up in North Cambridge, then two medical runs for people with respiratory conditions aggravated by the heat and humidity. The sky was darkening and threatening rain.

A little before four we received a call to go down to the Fenway Motor Inn, a modern twelve-story hotel down on the river between the Boston University and the River Street bridges. Some guy had fainted.

As we sped down Putnam Ave. fumes from the exhaust backed into the truck, making me feel even hotter and stickier than I had before.

"I wish I were beside a nice cool pool right now," I said.

"They got one at the hotel," said Joe.

"Maybe if we're lucky it'll be a drowning," Dave said, "and we'll have to jump in."

There was a bolt of lightning and then a crack of thunder. The truck braked hard. Some guy trying to turn left onto Western Ave. was holding up our lane of traffic. The air horn sounded loud enough to blast the cars right off the road, but they didn't move out of the way. Billy Stone pulled up over the curb to get around them. He was the best driver of us all; he had the smoothest touch with the apparatus and a natural instinct for getting through any kind of traffic.

At the hotel we grabbed the wind box and the chair and ran into the cool, air-conditioned lobby. A cop was holding the elevator for us.

"What's the problem?" Cooper asked him.

"I don't know," he said. "It looks like a heart attack."

Dave ran back out to the Rescue for the stretcher and the board, while we took the elevator up to the sixth floor.

There were more cops in the corridor. The room was down the hall to the left. The man lay at the foot of the bed. He was a big man, in his early thirties, and he was naked except for an undershirt. His face was blue. There was a smell of excrement, and the carpet was wet where he had voided.

A tall, attractive young woman rushed up to us. "Where's the Rescue?" she cried hysterically. "Where is it?"

"We're the Rescue," said Cooper, drawing her aside.

Joe and I fell over the guy. He was staring up at us with that nocturnal look, his pupils as big as an owl's. His ears, cheeks and lips were deep blue. All his vital functions had stopped. There was no pulse, no respiration: he was clinically dead, but he was still warm. I started cardiac massage immediately. Joe pulled an airway out of the wind box, stuck it in his mouth, placed the facepiece over his mouth, and began squeezing the ambu bag.

"Will he be all right? Please, tell me. What's wrong?" the woman cried to Cooper. She was dressed in a black cocktail dress, her blond hair piled high on her head. The bed behind her was unmade. Cooper told her the man had had a heart attack and she screamed, "Oh God, no," pressing her hands to her cheeks.

"How long has he been like this?" Cooper asked her quickly.

"Only a couple of minutes," she said. "He took a shower, then we just—then he started gasping."

"For how long?"

"I don't know, a minute maybe. Then he just stopped."

"Does he have a doctor here?"

"We're on a business trip. We don't know anyone here."

"All right. We'll take care of him," said Cooper, taking her arm and leading her from the room. He didn't want her to be watching; a resuscitation isn't a pretty sight.

The air conditioner was humming in the background as I kept up the massage, depressing the man's chest again and again.

"He's pinking up," said Joe. It was true. The blue was fading from the lobes of his ears. Billy Stone checked his pupils: they were starting to come down. We had a good resuscitation going.

Cooper tapped my shoulder and relieved me on the massage. Dave and one of the cops brought in the stretcher and the board and laid it down beside the guy. Billy Stone and I spread a sheet under him, then wrapped it around his waist and legs. When I looked up the woman was standing over us. A cop was right beside her.

"Is he all right?" she cried.

I got up and explained to her that she shouldn't be here. I asked the cop to take her down to the hospital.

"I want to be with him," she said. "Don't you understand?"

"You'll only get in the way," I said. "It's for his own good."

Dave, Billy Stone and I lifted him onto the stretcher. As we set him down the man suddenly gasped. We carried him out into the corridor, with Cooper and Joe keeping up the resuscitation all the while. The cops were holding the elevator; we had to tilt the stretcher to fit it inside. The doors closed and the elevator dropped. The sweat rolled down Cooper's face and his lips moved silently as he concentrated on the rhythm of the massage. We rushed the man out through the lobby. The woman was getting into the police car behind us as we lifted him into the Rescue. I shut the doors.

We pulled out. Dave had taken over the massage and Cooper was up front on the radio. The man drew another deep breath. He was definitely pinking up. The oxygen was getting through. I asked Dave if he needed a break, but he said he was all right. The guy was big, a strong man with sideburns and a great barrel chest. I checked his mouth to make sure it was clear; in cardiac cases we worry about vomit that might choke them. The guy drew another deep breath, then gasped again and again. Joe switched from rhythm to assist ventilation—trying to coordinate his efforts with the man's own breathing so that he wasn't forcing air in against him.

We pulled up to the emergency entrance of the Cambridge Hospital. The electric doors to the emergency ward were open. Inside I could see the hospital personnel: they had called a Code 99, assembling their resuscitation team. I relieved Dave on the massage. Billy Stone and Cooper opened the back doors and started lifting out the stretcher. Joe and I kept working as we moved into the hospital.

Out of the corner of my eye I could see doctors and nurses rushing ahead of us into the trauma room. They all lent a hand as we slipped him off the stretcher onto the table. Dave, Billy Stone and Cooper left immediately. I kept up the massage. One of the doctors motioned for Joe to stop ventilating. The doctor pulled out our airway and inserted a laryngoscope deep into the throat, then fed the endotracheal tube down into the guy and hooked it up to an ambu bag.

The place was full of people now and things were happening all at once. The doctors were shouting orders: to start an IV, to set up the EKG. One nurse stuck the IV needle into the man's forearm and taped it in place. Another nurse was pumping up the armband to take the blood pressure, while behind her an orderly rolled over the EKG machine. They quickly fastened the wire leads around the man's wrists and ankles. A fifth lead was secured by a suction cup right over his heart.

At the far end of the table a doctor was trying to draw blood from the man's calf for the blood-gas analysis. Without a pulse it was difficult to locate the femoral artery. The doctor jabbed twice, missing, then suddenly blood slapped across the table as the needle struck the artery, and the vial filled with red. At the same time another doctor plunged a huge horse needle of sodium bicarbonate directly into the man's stomach.

"Let up on the chest," a doctor ordered. I stopped the massage. He turned on the EKG machine, then held up the tape readout. I started the massage again.

"Ventricular fibrillation," the doctor shouted. "We're going to defibrillate."

Somebody moved the defibrillator into position. Somebody else squeezed jelly onto the underside of the discs. The doctor took the discs and rubbed them swiftly in circles over the man's chest. I stopped the massage.

"OK, everybody back from the table," the doctor shouted. He touched the discs to the man. I heard the electricity—zzzip—and saw the body jerk.

I started the massage again. They took another EKG reading. "He's still in VF," the doctor said. We backed away from the table and they hit him again with the defibrillator.

One of the doctors relieved me on the massage and I stood for a moment against the wall. The guy had definitely pinked up and looked a lot better than when we'd found him, but he was still a gruesome sight with all the wires attached to him, and the tubes and needles stuck into him. The floor was stained with blood and littered with piles of discarded EKG tape. The room smelled of alcohol and excrement.

"All right," said the first doctor, examining the latest EKG readout. "It's back. He's got a weak beat." They injected adrenaline directly into the heart with a cardiac needle. The anesthesiologist said the man was beginning to breathe on his own. I saw him gasp and draw a deep breath.

I went out into the corridor. In the waiting room his young wife was pacing nervously up and down. The clock on the wall read four-fourteen. Outside, the first heavy raindrops started to fall. Only a couple of minutes before we'd been sitting in the house watching the dark clouds pile up. It felt as if a lifetime had passed.

A few minutes later Mary Lane emerged from the room. "He's going on his own now," she said, hurrying toward the waiting room. "He's doing all right."

Whether or not he'd still be all right in a few hours, no one could tell. Whether or not there had been any brain damage wouldn't be known until later. All we knew as we ran out in the rain to the Rescue was that the man was still alive, that he had a chance and that we had played a part. He had been dead when we got to him, and now he was alive.

=*EIGHT* =

WE'D JUST STARTED IN ON THE MORNING housework when we had a call from an address on Dana Street. When we got there we rushed upstairs to a second-floor apartment. A young guy in pajamas opened the door partway and scanned our faces. Cooper asked him what the problem was.

"Problem?" the kid said. He had a nervous smile that flashed on and off. "Oh, there's no big problem. I mean there's just a little problem, really. With my wife. She kind of hurt her back."

"Well, are you going to let us in?" Billy said.

"In?" said the kid, glancing back over his shoulder. "Oh, sure. Come on in if you want."

He opened the door. The girl was right there in the front room, on a mattress laid directly on the floor. She was a small, pretty girl with a pixie haircut. Her eyes leapt to our faces, but the body under the blankets remained rigid as a mummy. Except for a few boxes and suitcases along the walls the room was bare.

Cooper asked her what had happened.

She shot a glance at the kid. He was a tall, fair-com-plexioned kid, and before our eyes he turned bright pink.

"Well," he began, gesturing with both hands. "It's ac-tually very simple. It happened last night. Amy thought we could—I mean she said that . . . " The kid stopped suddenly. "Do you really have to know exactly how it happened? We only got married yesterday."

"I see," said Billy. He brushed the hair back off his forehead. "Well, ma'am" he said, turning to the girl, "can you move your legs?"

"Oh, sure she can," the kid said, kneeling down by her side. "She moves them just fine. It's not that serious at all."

"It hurts when I move them," the girl said, ignoring him, "and it hurts when I lift my head. It hurts all over."

Cooper leaned over the mattress and had the girl squeeze his hand with each of her hands. She seemed all right. Billy Stone went down for the orthopedic stretcher.

"Well, just make yourselves at home," the kid told us while we waited. There was no place to sit so we remained standing in the middle of the room with our hands in our pockets. The kid grinned at us, and we all grinned back at him. "Nice day," he said to Dave.

"Oh, it's a beautiful day," said Davey back. The poor kid was clearly unnerved. He'd probably been expecting two guys in white coats to take the girl quietly out of the house. Instead he'd gotten sirens, a fire truck and five brutes in blue storming through his door.

We ran her down to the hospital, then rode slowly back toward the house, crawling along in the heavy morning traffic. It was a bright September day and Harvard was back in session so there was a steady stream of bicycles flowing along both sides of the street. A kid pedaling an expensive ten-speeder just behind us was wearing a huge steel chain and lock around his neck. It struck me as absurd; man chained by his possessions. I remember reading somewhere once that Cambridge ranks number one in the nation in car thefts per capita. I don't know whether that's true, but I wouldn't ever be surprised to learn that we rank first in the universe in bicycle thefts.

"Hey, Billy," Dave said, "what ever happened to that ten-speeder you were gonna buy?"

"What happened to it was that I bought it and then I was afraid to use the damn thing," Billy said. "So finally I rode it down to a union meeting one night. I figured no one would dare to rip it off over there. I had a big chain lock and I connected it to an iron grating inside the hallway. When I come out the bike was gone. That's what happened to it. It's brutal out there."

"You're telling me," Dave said. "It's like the Wild West. It's gotten so bad the cops just figure that everybody's riding around on stolen property. You remember the red bike I used to have?"

"That old piece of junk," Billy said.

"That's the one," Dave said. "It was a real nothing, right? Well, one night I rode it over to Kelly's just for the hell of it, and we had a couple of beers together, and, Jeez, I come out of there feeling no pain and it was a beautiful night. So I'm riding through Harvard Square, singing at the top of my lungs, when this cop flags me down. He struts up to me, you know the way some of them do in the Square, like they were the last badge between here and Dodge City. This guy's new and he doesn't know me.

" 'Is that your bike?' he says.

" 'Sure it's my bike.'

" 'Can you prove it?'

" 'Sure I can prove it. I'm riding it, ain't I?'

" 'You got a bill of sale for that bike? You got some identification? How do I know you didn't steal it?'

" '*Steal it?*' I said. 'Look at me. You think I'd steal a piece of shit like this?'

"He didn't answer me. He made me give him my driver's license. There was an old license plate on the bike and he wrote that down, too.

" 'I'm gonna check up on this,' he says.

" 'Go ahead,' I said. 'Check up.'

" 'You're drunk,' he said.

" 'Yeah, well,' I said, 'let's just say I'm pleasantly intoxicated, officer.'

" 'All right,' he says, sticking a finger at me. 'Now you listen to me and you listen good. You get the hell outta here. And don't let me catch you around here again.'

" '*Get the hell outta here?*' I says. 'Whaddaya mean get the hell outta here? I *live* here!'

"Well, just don't let me catch you around here,' he says.

" 'You've gotta be kidding,' I said. 'I've lived here all my life. I got as much right as anybody to be here.'

"He spins around with both fists on his belt. 'Get the fuck outta here,' he shouts, 'before I run you in for drunken driving.'

" 'Oh brother,' I said to myself, and I pedaled out of there as fast as a bastard. This place is crawling with geniuses and I have to run into the only donkey this side of the Mississippi. I rode that piece of shit home, parked it in the back of the garage and I haven't touched it since."

"You're lucky he didn't lock you up, Dave," said Billy. "Granatino would've locked you up."

"You're telling me," Dave laughed. "And if he wasn't one of my best friends he'd do a lot worse than that."

Back at quarters the other companies were way ahead of us on the housework. We still had all of the second floor to do: the company dormitories, the officers' dormitory, the Chief and the Deputy Chief's room and the lavatories, as well as windows to clean and poles to polish. Then we'd have to clean the Rescue inside and out and complete an inventory of medical equipment aboard. And after that we'd have to drill on our procedures. The housework usually takes about three hours. Even if we didn't get another call all morning we'd be lucky to get through the chores by noon.

Dave and Billy Stone took the bathrooms; Finnegan and I, the dormitories. I swept floors and emptied wastebaskets while Joe trailed along behind with a mop and a steaming bucket of hot soapy water. From out in the hallway came the sound of cheerful whistling, and Kilroy appeared in the doorway, a mop of his own slung over his shoulder, and sang us a chorus of "Whistle While You Work."

"Well, if it isn't one of the original seven dwarfs," said Finnegan, wringing out his mop.

"That's right, Joe sweetheart," Eddie said, "and this is a day you'll never forget. You know, Joseph, when you stop to think about it, it's incredible. Here we are living in the space age, but we're still using stone-age methods to clean our floors."

Joe broke into a huge grin. "Aw, Christ," he said, bending over the bucket of water and turning away.

"Joe, I'm gonna ignore that remark," said Kilroy, "and share with you the discovery of a lifetime. A buddy of mine, presently employed as a janitor in a chemistry laboratory run by a prominent professor who has recently returned from the Amazon, has just sold me, strictly under the counter, a sample of a new, secret substance."

"What substance?" Joe said, pretending to scratch his mustache in order to hide a smile of disbelief. "I never heard of any secret substance."

"Of course you haven't. It's a top secret that could revolutionize the entire cleaning industry. Excuse me, Larry," Kilroy said, relieving me of a wastebasket that was half full of the dirt I'd swept up. He dumped the contents of the basket onto the floor.

"Jesus," shouted Joe. "What the Christ're you doing? We're trying to clean up this goddamn place."

"Relax," said Kilroy, smearing the dirt around with his feet. "All I'm asking for is a minute of your time that could save you literally hundreds of hours of work." He examined the mess critically, then traced a line up the middle with the heel of his foot. "Now Joe, you mop your side with your everyday cleaner and I'll mop mine with the new miracle ingredient."

Finnegan's face was red. "Where's this new miracle ingredient?" he said unhappily. "I don't see no ingredient."

"It's already on my mop," Kilroy said. "A little goes a long way. Ready?"

"Aw, Christ," Joe said, turning his back to Kilroy. He took a halfhearted swipe at the filth with his mop, then started mopping in earnest. In half a minute he had turned his side of the floor into a pool of mud.

"OK. Not too good, right?" said Kilroy. "Now watch this. I won't even look." Trailing his mop behind him with one hand, Kilroy walked through the muck straight to the door.

"Well?" he said from the doorway.

"Well what?" shouted Joe. "I don't see any goddamn difference."

"You don't? Jesus," Kilroy said, scratching his head, "it worked all right downstairs. Hey," he said suddenly, "you don't suppose that guy sold me a fast bill of goods? I'm gonna give that sonofabitch a call right now before he takes off. Quick. Have you got a dime?"

"Jeez, I don't know," said Joe, fishing through his pockets. "I think all I've got is a quarter."

"I'll take it," Kilroy said. "I'm gonna give that bastard a piece of my mind. See ya." Then he was gone.

I stood there staring dumbfounded at Finnegan. Joe looked at the floor, then at me, and then, throwing down his mop he collapsed on one of the beds, howling and shrieking in a fit of laughter.

"What the hell's so funny?" I yelled. "How could you have fallen for that? You stupid ass! Now we've got to clean up this mess."

Joe didn't answer. He couldn't stop laughing. He slapped at the bed, the tears rolling down his cheeks.

"He didn't even remember," Joe gasped, still doubled up with laughter. "He pulled that exact same bit on me four years ago at Engine Six and he didn't even remember. I *love* that bit."

"Well, then you can clean it up," I said, heading for the door.

"Sure. Sure," he said, waving a hand at me. I left him in there rolling on the bed and laughing so hard he had to hold his sides. I don't think he even noticed me go.

About ten minutes later we had a call: an injured person in an apartment building near Inman Square. Joe, now in a more sober mood, was at the wheel as we raced up Hampshire Street. The morning was so brilliantly clear we could see every sparkling window on the skyscrapers in downtown Boston.

"Need anything from the pharmacy?" Dave asked as we shot past a corner drugstore.

I smiled. We'd had a fire there two years ago. Davey fell through a trapdoor into the burning cellar, and when I got to him his mask was fouled up and he couldn't breathe. He held on to me and I managed to get him up the stairs; then Cooper helped me take him out.

When we got to the building a young guy met us at the front door and led us down the corridor to a rear apartment. "He's in the bedroom," said the kid. We rushed through the lavishly furnished living room into the bedroom. Another young guy lay on the bed clutching to his chin a towel that was sopping with blood. The sheets, pillows and blankets were all smeared with blood. His fingers were dark red but his face was ghost-white.

Cooper pried loose the guy's grip on the towel. As he lifted it gently a clot of blood the size of a softball slid down the side of the kid's neck. His throat was slit wide open. When he raised his eyes and tried to speak it looked as though he had two mouths. Cooper took a clean towel and wrapped it around the kid's throat, just tight enough to keep pressure on the wound. Then we lifted him into the chair and rushed him out to the Rescue. The other kid came aboard with us.

On the way down we worried about the kid's going into shock. He was very pale and had lost a lot of blood. We covered him well and he lay very still while Dave and Billy Stone kept talking to him, reassuring him, telling him he was going to be all right. We all had blood on us except for the other kid, who sat in the rear casually inspecting the contents of the Rescue as though attempted suicides were everyday occurrences for him.

We rushed the kid into the Emergency Room at Cambridge Hospital and lifted him onto a bed. The nurses and interns swarmed around him. We washed our hands at the sink, and as we were starting out of the room he suddenly spoke. "Everybody wants to fuck me," he said, "and that guy right there fucked me."

Standing in the doorway was the kid who had ridden down with us. He shouldn't have been in the Emergency Room at all, but in the confusion of our arrival he had been overlooked. For a moment we were all too embarrassed to move. It's one thing to joke about homosexuality out on the street; but in here, under these circumstances, it set us all back.

"I'm sorry," said Cooper as we moved toward the door. "You're not supposed to be in here." Billy led the guy over to the waiting room.

"Didja hear that?" the guy said, loud enough for us all to hear. "He says I fucked him, and that's why he cut his throat.

That's a lie. He just couldn't take it is why. He never could. I shouldn't have even bothered calling you."

"All right," said Billy, opening the door. "Why don't you just wait out there."

The guy hesitated. "Didja hear what he said?" he shouted.

"Yeah," said Billy. "We all heard it. More than we wanted to."

Nobody said much in the truck on the trip back to quarters. The other companies were already out on the apron, washing down their apparatus.

"Hey, what's with the long faces?" Kilroy shouted as we climbed down from the Rescue.

"Hey, Joe," called Jerry Martin from the side of the Aerial Tower, holding up a can of grease. "I wonder if I could interest you in trying a new kind of polish." They all cracked up.

We trudged upstairs to our housework. "Jesus," Dave said as we climbed, "call me a cab so I can get out of this madhouse."

No one responded.

"Jesus," said Dave, "isn't anybody gonna call me a cab?"

"All right," Finnegan said. "You're a cab."

"Thank you," said Dave. "That's more like it."

We split up to do our separate tasks.

It was already nine-thirty and we still had a mountain of housework, but we hadn't made much of a dent in it when the bells started coming in. A whole lot of bells. We hit the poles. I counted the bongs. Box three-nine-two-four. Everything was going.

We pulled out and raced up Broadway with our siren screaming and the Deputy's car, the pumper, the hose wagon and the Tower strung out behind us.

The radio crackled.

"Fire Alarm broadcasting an alarm of fire box three-nine-two-four at the corner of Pearl and Valentine streets."

"Car Two to Fire Alarm. Message received. Box three-nine-two-four. Pearl and Valentine."

Billy Stone and I began pulling on our coats. "It's false," said Dave, unmoving. "It's too nice a day for a fire." The air brakes hissed as the truck slowed, then wheeled sharply to the right and sped down Prospect Street with the sirens echoing off the houses.

"Engine Six to Fire Alarm."

"Engine Six."

"Fire showing."

Now Dave jumped up, and quickly we all pulled on oxygen tanks, boots and gloves. The air horn blasted and we burst into Central Square, careening past stalled traffic, then made a sharp right. As we roared up Pearl Street we could smell it. It smelled strange; not the familiar stench of burning wood and plaster, but something far more pungent.

"I'll *say* there's fire showing!" Cooper yelled.

"Holy Jesus," said Billy Stone, peering forward. To the left, above the roofs of the houses flashing past us, a huge column of dense black smoke rose over the city. Straight ahead we saw the red and white lights of other apparatus. An excited voice came over the radio.

"Engine Six to Fire Alarm."

"Engine Six."

"On the order of Lieutenant Druhan give us a second alarm."

"Oh brother," said Davey, handing Billy Stone a halligan bar and grabbing an ax for himself as the second alarm began sounding over the radio.

Joe turned onto Valentine Street, pulling up at the right-hand corner alongside a store fronting on Pearl. We leapt from the Rescue. Black smoke was spewing from all the doors and windows of a two-story brick plastics warehouse. The surrounding residential area was tightly packed with rows of old wooden three-decker houses. Directly ahead of us in the smoke-filled street were Engines Five and Six. A couple of lines had already been laid.

The stench of the smoke was overpowering. Dave, Billy Stone and I pulled on our masks and followed one of the lines into a narrow alleyway formed by the near side of the warehouse and the rear of the corner store. Ahead of us a cellar door belched thick black smoke. We followed the line down four steps, through a door, smack into the blackness and the heat. The fire encircled us. Over the furious roar we could just barely make out the sound of coughing. We moved in after them, keeping the line between our feet to guide us. They were just a little way in from the door: it was Harry Sloan and Engine Six's crew. The line was going. They were all choking. Without masks I couldn't believe they had made it even that far in the hellish scalding-hot smoke. They were taking a tremendous beating. Over the fire I could hear Davey yelling for them to get out and get masks.

We took the heavy line from them and started moving farther into the cellar, Davey on the tip, me backing him up and Billy Stone jockeying the line. I kept close against Davey, cradling the fat hose in my arms and leaning forward into the back pressure of the line. The water sizzled and crackled as it exploded through the tip. Cold spray rebounded all over us. The red was straight ahead.

"I'm gonna hit the ceiling now," Dave shouted, and together we raised the tip. The stream crashed directly in front, lashing back at us. We weren't hitting the red. We took a step forward and slammed into something. I reached out into the darkness. We were surrounded by a forest of man-size rolls of plastic that blocked our advance and our line of fire.

"I'm gonna try to hit it over to the left," Dave yelled. "Let's move it over to the left." I shouted to Billy to take up on the line. The line slackened and we worked our way in among the rolls of material, with the roar of the fire deafening, and the heat growing unbearable. The fire was all around us, but we couldn't seem to get a good shot at it. We were hitting it, but we weren't knocking any of it down. Either there was more stockpiled material we couldn't see blocking our line of fire, or we were hitting the reflection of the fire off the ceiling and the plastic rather than the fire itself. It could have been either or it could have been both. It hardly mattered. All we knew was that we were surrounded by red, that we were hitting it, but that none of it was darkening down. It was like a nightmare in which the gun jams in your hand and your legs won't carry you away.

The smoke was so thick I could smell the burning plastic right through my mask. I pulled the strap as tight as I could to shut out the stinking fumes. Flames shot up just to the right of us. I yelled at Dave and we turned the tip and knocked them down. Then there were more flames dancing on the other side of us. The murderous heat radiating from every direction burned my ears and the back of my neck. It was like an inferno.

Then it happened; all of a sudden Dave was shouting. Fire flared above us like the crest of an enormous wave. The dark exploded in a searing blast and we were being driven backward in a sea of fire that sucked me down. An image of my son waving good-bye flashed through my mind, and I fought my way to my feet. Billy Stone grabbed hold of me and Dave was screaming to keep the tip up. Dragging the heavy line with us, we struggled to keep the tip pointed at the ceiling to make a shield of water around us. We hardly seemed to be moving. The blood hammered hard in my head. My lungs felt as if they were going to burst. I drew in deep on the air but couldn't get as much as I wanted through the mask.

We made it to the door and retreated out to the base of the stairs. Just as suddenly as the wave of fire had mounted and broken, it was spent. We repositioned ourselves, then began moving again into the cellar, pushing the fire back along the ceiling. We got in as far as we had been before and then some. I heard a sharp creaking of timbers. Water was raining down from overhead, and I started to worry about what was happening up on the first floor. The ceiling had to have been weakened by the tremendous amount of fire against it. Because of the stockpiled material, we had to move deep into the cellar to get a crack at the seat of the fire; but the farther in we went, the more we exposed ourselves to the danger of the ceiling's collapsing. The fire was just too intense to run that risk.

I grabbed hold of Dave and Billy. "Don't move in any farther," I shouted. "I'm going upstairs to see what's going on overhead. Stay right here."

I made it to the door and hustled through the alley to the front of the building, the daylight momentarily blinding me. The street was packed now with apparatus. Ladder Three had thrown its stick, the Aerial Tower was pulling into position and more apparatus was maneuvering out on Pearl Street. Lines were strewn everywhere, running through doors and windows into the warehouse. Power lines dangled and the tip of one telephone pole was burning, showering sparks like a Roman candle.

I scrambled up onto the loading platform and started to follow the lines inside as Jerry Martin and the Lieutenant from Engine Five emerged from the smoke.

"Jerry, how's the floor in there?" I shouted.

"It's ready to go," Jerry yelled, coughing badly. "We're backing out now."

I ran back down into the cellar; Dave and Billy weren't

where I had left them. They had veered off to the right. Keeping the hard line between my boots, I moved in after them, yelling as I went but hearing nothing over the roar of the fire. Water was pouring down through the ceiling now. Underneath my coat I was swimming in sweat. My head pounded from the tight straps, but the stink of the burning plastic still invaded my mask.

I slammed into someone. "Hey, take it easy, will ya?" Dave shouted.

"We've got to back the hell out of here," I yelled. "The ceiling's going to go."

Dave turned to Billy, who held the tip. "We've got to make a run for the roses," Dave shouted to him, and then we were backing out quickly, hauling the line with us all the way outside to the base of the steps, where we prepared to make a stand.

But we couldn't hit it very well from outside, and the fire quickly advanced. In a few minutes flames were lapping over the door itself, forcing us back up the steps. It was really tremendous now; we were hitting it point-blank but we weren't knocking down any of it. A two-and-a-half-inch line is a big line, but in the face of that wall of flame we might just as well have been armed with water pistols.

From deep inside the cellar came a great thunderous rumble. We hit the deck and tongues of flame shot out over our heads. Fire rocketed skyward from the doorway, whipping and snapping at us as we crawled away from it, dragging the line with us across the smoke-filled alley.

Somebody—the Deputy—grabbed hold of my shoulder. "What's the situation here?" he asked. I told him that we believed part of the ceiling had fallen and that there was stockpiled material in the cellar blocking our advance.

"OK," the Deputy said. "If this breaks out of here we've lost the building." He told us that the third alarm had been sounded, and that he wanted us to hold this position. Then he was gone.

We began to move back in on the fire. Even if the third alarm had been sounded, it would still be a while before the off-duty shift got here. All the apparatus was probably already on the scene, but there's only so much apparatus you can squeeze into a given area, only so many hydrants you can draw water from, before it becomes a question of manpower and luck. The area was so tightly packed with tinderbox housing that if the fire did break out of the building, or if a strong wind suddenly came up, we could have a conflagration on our hands. But those were only ifs and the building wasn't lost yet. As far as we knew the fire was still confined to the cellar and the first floor. But it was intensifying rapidly, and unless somebody could get water on the seat of it from another position, it could take off. One thing was clear from our end: it didn't look good. Fire fighting takes place inside. If you're going to save a building you can't do it from the outside. And when the fire pushes you out, you have to fight to get back in.

We worked our way back in close to the fire. I held the tip now, directing the stream into the center of the blazing doorway and playing it up on the wall where the flames overlapped. A tremendous volume of fire was shooting out in awesome, blinding sheets of blue, green and purple. I heard distant bells ringing. They were the warning bells on my air tank: the tank was nearly empty. Davey slapped my shoulder and took the tip. I ran out to the Rescue to change tanks. The Rescue had been backed out onto Pearl Street and with all the apparatus out there it took me half a minute to locate it. When I returned to our position Billy's tank was low. When he got back, Dave went.

We held on to that position for nearly half an hour, but we were unable to get back inside and the fire gradually gained on us. Jets of flame forced us all the way back against the rear of the stores. Waves of fire scurried up the wall over the bulging bricks. A steady spray rained down on us, and looking up through the shifting smoke I realized with surprise that there were other lines going from the roofs behind us.

The Deputy's aide appeared from out of the smoke. "They want you outta here," he shouted. "The wall's gonna go."

We hauled the line out of the alleyway into the street, where the Deputy himself was waiting.

"Get that line up there," he said, pointing to the roof of the corner store. A ladder had already been thrown against the side of the building. We took off our tanks and masks and left them on the sidewalk; then, throwing the stiff charged line over our shoulders, we climbed up to the flat tar-and-gravel roof.

It was our first glimpse of the scale of the battle. Directly in front of us solid sheets of flame enveloped the entire side of the warehouse. Overhead a bubbling, churning mass of orange flame had broken through the top of the roof and great clouds of black smoke rolled skyward. A funnel-shaped cloud of smoke towered hundreds of feet in the air, drifting slowly back over the city. A helicopter flew under it, its thrashing blade barely audible above the roar of the fire and the sound of our lines.

We were hitting it from every side now: from the deck guns of the wagons positioned out in the street, from the raised bucket of the Aerial Tower, from the roofs of the stores farther down to our right and from the yards and porches of houses behind the warehouse. We were hitting it from all over, but the monstrous body of smoke and fire just swallowed the streams of water thirstily.

Exhausted from nearly a full hour of wrestling with the heavy line, Dave, Billy and I took up a position at the rear of the roof and settled in for a long siege. Since we weren't wearing the masks, the shifting smoke quickly made our eyes water. My throat felt dry and I began to cough. We braced ourselves against the line, opened up the nozzle and started hitting the side of the warehouse. Where the wall bulged, we could see that bricks and crumbling mortar were falling down into the alley. Davey turned to me, bleary-eyed, his face blackened with soot, his mouth and nose running with dark mucus, and smiled. He gestured toward my head and he and Billy broke out laughing.

"What's so funny?" I asked.

"Look at your helmet," Billy said.

I was much too tired to remove my helmet for a joke. But I didn't have to because, looking at theirs, I saw what they meant. The once-red helmets were charred and bubbly. The heat in the cellar when the fire broke over our heads had blistered the paint right off.

Dave and Billy thought it was funny. I was too hot and exhausted to be amused. My throat was sore and my head ached. I would have given anything to be able to relax with a cool drink. . . .

Only a couple of days before, Bev and I had been to a cocktail party. It was given by the Commander of my reserve unit, who owned a big house over in Newton with a spacious lawn and a back terrace that overlooked the garden. The Commander was a doctor in civilian life, like most of the officers in our hospital unit, and he threw a good party. The place was filled with people, perfume and the loud chatter of conversation. The drinks were beginning to warm my stomach and I had that pleasant feeling of leaving behind the day-to-day grind that you get when a party starts going well.

But then the wife of the guy I'd been talking to turned to me, and it was the same old bit all over again. She was a tall, handsome woman, wearing an elegant long dress and a few thousand bucks' worth of rocks around her throat. Her husband was a doctor out in Concord.

"You'll have to excuse me," she said. "My memory's simply awful lately. Where did you say you and your lovely wife are living?"

"Over in Cambridge," I said.

"Over in Cambridge, oh how nice," she said. "I sometimes wish we still lived there. It's so dreadfully calm in the country, you know. And do you practice there also?"

"In a manner of speaking," I said.

"Oh, really? Where?"

"Up at the main firehouse."

She smiled. "I'm afraid I don't understand," she said. "Do you run some sort of clinic there?"

"No. I'm not a doctor," I said. "I'm a fireman."

"A *what?*"

"A fireman."

She looked at me as though I'd just spilled my drink down her dress. "Oh, I see!" she said gaily, trying to make up for the look of astonishment. "Oh, you mean you go through the streets with the red trucks and all. That must be very exciting work." She cast a quick nervous glance in the direction of her husband. She smiled warmly at me, but the look in her eyes told me she might just as well be talking with the mad bomber as with a grown man who rode around on fire trucks.

I let it drop. What is there to say: "I'm sorry you have the wrong impression of what fire fighters do. It's really difficult, demanding, dirty, dangerous work that I nevertheless find deeply rewarding"?

I think fire fighting is one of the most underrated and misunderstood occupations in this country. Most people don't have any idea of what we do, and in large part it's our own fault: how can they know if we don't tell them?

I remember once attending a union meeting where

members were beefing about our continual lack of success in
getting the city to vote us the benefits and compensations
they felt the job merited. I said, "Well, why? Why should we
get any of that?" They said, "Look at what we've got to do."
I pointed out that the only people who hold fire fighting in
high esteem are firemen themselves. Most people ride past a
firehouse and it seems to them we don't do anything between
fires.

It's particularly hard for the public to understand the
monetary value of what we do. A fully trained and equipped
fire department represents a huge item in any city budget,
and doesn't seem to produce anything of tangible value. Of
course a fire department is a form of insurance, the worth of
which a town should calculate in terms of property and jobs
saved, but how do you begin to assess the worth of a conflagration prevented or a life saved?

Local governments are also reluctant to give firemen
special benefits, because they're afraid if they do they will be
besieged by other groups with demands for equal consider-
ation. Now, I'm all for having garbage men and other munici-
pal employees earn as much as possible, but I think it should
be recognized that fire fighting involves a level of
commitment and risk which entitles firemen to special con-
sideration. Modern fire fighting is a complex, highly technical operation and a tremendous amount of time is taken up
with training and maintaining a state of readiness, but people don't see this.

Across the entire United States firemen need an up-to-
date public relations effort to erase demeaning stereotypes.
Kids have traditionally been taught to think of firemen as
friendly old men in suspenders who fix bicycles and rescue an
occasional cat. TV commercials invariably present firemen
in a comical light, just as scenes in the old-time movies
showed firemen racing every which way without managing
to get water on anything but themselves. Nowhere is anyone

ever told or shown what it's actually like inside a fire and what's involved in combating it. The public image of us remains a comical one. Comical, that is, until we're needed.

I remember one fire at a construction site on Mass. Ave. An acetylene tank had exploded, causing a small fire in the subcellar. When we arrived, the workmen on the girders seven floors above ground level taunted us. "Hey, firemen," they yelled. "Help. Save us." We heard them laughing. It was all a big joke. Suddenly there was a series of explosions and a ball of fire shot across the underside of the first floor. The stockpile of over one hundred tanks began exploding one by one. The hard hats slid all the way down the I-beams and took off. We had to go into the fire and remove the unexploded tanks by hand. They were warm and could have blown up in our faces at any minute.

The point is not that all firemen are heroes. Firemen are ordinary guys who are dedicated to a job which may demand on any given day that they risk their lives for the good of the community. And every fireman, whether or not he'll admit it, is prepared to do just that, because he believes in and is proud of the work he does. This fire in the plastics warehouse was a case in point. In the early minutes, when we were inside fighting at close quarters, there had been great danger from the toxic fumes and the possibility of structural collapse. But no one had hesitated to carry out the necessary tasks.

Now, however, from our vantage point on top of the roof, it appeared that all companies were safely outside. The building itself was fully enveloped in flame and the battle was simply to contain the fire. Word came that the Deputy needed all available backup men to help cover the exposures behind the warehouse. I left Dave and Billy with our line and made my way to the back of the warehouse, where I helped lay another line. There were at least half a dozen old wooden three-family houses almost flush against the rear of the blazing building. The people had been evacuated and a battery of lines had been assembled in driveways, in yards, on roofs and on porches to form a water curtain that would protect the houses.

It was a constant battle to fend off the flames. At one point fire caught hold of part of a garage; a crew quickly rushed in to put it out. Minutes later flames began lapping up against a second-story porch; they, too, were immediately knocked down. Because of the size and intensity of the fire and the proximity and flammability of the houses, it took a tremendous effort from all hands to keep it contained.

For the next couple of hours we went on laying lines, jockeying lines, taking turns on the tip. We took our breaks right on the lines, drinking coffee where we stood. But with all the extra off-duty personnel that were coming in, we couldn't understand why we weren't getting at least some time away from the lines.

Just after two o'clock a section of the roof collapsed. I felt the ground move; then a column of flame, fueled by the fresh oxygen, rocketed a hundred feet into the air. The wind shifted slightly and a huge mushrooming cloud of black smoke obscured the sun. Sparks and burning embers began raining down from the darkened sky, and people were shouting.

I ran around to the front of the building and saw the Aerial Tower bucket being lowered. Inside, Jerry Martin and John Riordan were bent over, shielding their faces. They had been suspended high above the street when the swirling tentacles of thick black smoke had engulfed them.

Now Bob Mondello and Ken Hovey climbed up to help them down. They were choking and gasping. Jerry Martin shook them off and staggered to the gutter, where he doubled over as a stream of black vomit gushed from between his lips. Watching him made me feel sick myself. The foul-smelling smoke was everywhere now, and there was no way to avoid it. It stung my eyes and burned my throat, which was so raw now that it hurt each time I swallowed.

Eddie Kilroy was standing by Engine Three, ankle-deep in water. The surface of the flooded street reflected the glow of the fire.

"We'll be here all fuckin' week," he said. "They'll have to get a crane and pull it apart. That junk in there will burn like charcoal."

"What I can't understand," I said, "is why we're not getting more help."

"It's the smoke," Kilroy told me. "A lot of guys have been overcome. They just ran a whole bunch down to the hospital. The guys in the first-alarm assignment have been throwin' up all over the streets."

I helped Kilroy relocate one of Engine Three's lines to the front of the building. We stepped over the web of criss-crossing lines that lay half-submerged in the flooded streets, like snakes. On the wagons dog-tired men leaned against the deck guns, directing their feeble streams of water into the fire. The sound of water rushing from all the lines made me sleepy. I had a splitting headache. Most of all I felt exhausted.

Outdoor fires are always exhausting. They're tough because you're doing heavy work all the time, and because you can't use the masks and the smoke gets to you. Room fires can be much more dangerous, but at least they're over fast and don't wear you down as much.

Still, the most exhausting tour of duty I ever served wasn't a fire. It was simply my first night on the Department. I had no idea what I was walking into. I had passed the exams, but I had absolutely no practical knowledge of what fire fighting involved. In a small city like Cambridge it wasn't feasible to set up a drill school for one man, so until there were enough of us to form a class, I was to receive my training on the job.

I reported that evening to the Lieutenant at Ladder Three. He wasn't too impressed with my sales background or my total lack of experience. He introduced me to his crew, told me what my duties would be and issued me bedclothes. Then he assigned me a partner at whose side I was to remain at all times and from whom I was to take orders.

I put my rubber goods on the truck, sat down on the rear step and tried to make myself at home. Some guy came up to me and introduced himself as the fireman who ran the busiest library in Cambridge. He said his name was Jerry Martin, and he insisted I should come up to his locker to see his books.

"But suppose there's an alarm?" I asked as we climbed to the second floor.

"What're you, nuts?" he said. "You could easily go three or four hours without an alarm."

Suddenly the lights flashed on and bells began clanging. I froze on the spot. "Is that for me?" I asked.

"Hell, yes," he shouted. "What the hell are you waiting for?"

I slid the pole and jumped onto the truck, clutching the side bar for dear life with one hand and holding onto my helmet with the other. My coat was flying in the wind as we raced through the streets at what felt like a hundred miles an hour. I was deafened by the sirens, blinded by the lights and in terror of falling off. I had no idea where I was or what I was doing. When the truck finally stopped, I tumbled down onto solid ground and realized all at once that it was only a false alarm, that I was standing just two blocks from where I had grown up and that I had my coat clipped up unevenly.

Back at the house everyone went up to bed around midnight. I decided it would be a good idea to sleep with my clothes on. In the dark I saw one man kneeling by his bed and praying. Although I hadn't prayed in years, I suddenly thought it wouldn't be a bad idea. I prayed that I would be allowed to get through the night.

I couldn't sleep at all. I was afraid that if I fell asleep I'd miss any alarms. I decided to go down to the apparatus floor and sleep on the back of the ladder truck. While I was sitting on the rear step I heard a loud bong. I sprang to my feet and began pulling on my coat and helmet, careful to get them on right this time. I was the first one on the truck and I waited proudly at my post for my fellow crew members to come down and see how ready and alert I was and how much they could count on me. But no one came down the poles. After a minute I climbed down off the truck and saw that the wall clock at the rear of the floor read two A.M. The bong I'd heard had marked the hour.

I sat down again on the step and tried to sleep, but I was up for the bongs at three, four, five and six. I managed to survive that first night, but the next morning I would have bet a month of fat sales commissions that I wouldn't survive a week of it.

Everything that happened in those first weeks increased my personal admiration for the men I was working with, and at the same time heightened my sense of my own incompetence. One morning at four-thirty an alarm came in for a fire at Harvard's Quincy House. When we got there smoke was pouring out of the windows on the fourth floor. They raised the ladder, and members of my company made their way up to the roof. One of the firemen yelled down to me from the fourth floor to bring him a Scott Airpak. I knew that the Airpaks were tanklike cylinders of compressed air, but I wasn't sure where they were kept. Luckily I found them in a black case, and I started up the ladder with one.

"Hey, you," he hollered from the fourth floor. "Are you a fireman or a salesman?"

"What the hell's that supposed to mean?" I shouted back.

"You look like a goddamn tie salesman climbing up with

that case. You're supposed to take the cylinder out and put it on your back."

I went back down, put on the Airpak haphazardly and climbed back up. While I was up there the Lieutenant of the Rescue Company came to the window and shouted had anyone seen any of his crew. Another Lieutenant was yelling for us to evacuate. The Engine Companies, hitting the fire from the other side, were driving the flames in our direction.

We climbed down and I looked on as they began to maneuver the ladder from the roof to the fourth-floor window. I could see the glow of the fire behind them as they climbed out onto the window ledge and hung by their hands. Just before the ladder was properly positioned one of them decided to risk it all, and took a daring leap. He made it. The others waited a few seconds more, then managed to grab hold of the ladder and scramble down to safety. Standing in the middle of the students, who were cheering their escape, I felt like just another bystander.

At the end of that first month I began to think seriously of going back to sales. Aside from the embarrassments of on-the-job training, there was the continual feeling of being a bumbling outsider who endangered the performance of a first-class crew by his lack of knowledge. I began to wonder whether I was really cut out for this life. What kind of men, I wondered, would be willing to risk their lives day in and day out?

I tried to figure out their motivations, and concluded only that theirs had to be different from mine. In comparison with the other men I had lived a soft life. Even my years in the Reserves hadn't prepared me for this. I'd been an officer in the infantry but I had never been to war.

I managed to hang on thanks to one guy, my partner, Jim Finley. He guided and encouraged me. He made me realize that I was learning an entirely new profession, that there was much more to fire fighting than just climbing a ladder and aiming a hose, and that a man just couldn't walk in off the street and do the job. He said it was a question of training and experience and that I'd pick it up eventually. He had been on the Rescue and he told me about what he'd done and learned there. He said the thing for me to do was to try to get on the Rescue after drill school. By his personal example and his help Jim Finley made me want to hang on until I could hold up my own end.

I wasn't holding up my end now. I was exhausted from fighting the fire and sickened by the noxious fumes. I no longer had the strength to haul line or take my turn on the tip. I sat on the curb with my boots in inches of black, ash-stinking water, feeling too weak to hold my head up straight. At least it didn't matter much now. The peak of the fire had passed. Fire glowed inside the smoking ruins of the building, but most of the flames had died. Individual firemen trailing inch-and-a-half lines worked their way in close to the smoldering rubble, finishing off isolated pockets of fire that burned behind the jagged outline of half-crumbled walls.

The throbbing of a nearby pumper sent waves of nausea through me. I lowered my head between my knees and tried not to think about the smell of the smoke, but it was everywhere, in my clothes, my nostrils, my hands. My chest hurt when I breathed and I couldn't seem to get my wind back. Davey came sloshing through the black water and the ripples made my head spin. He helped me to my feet and we made our way slowly back to the Rescue. Cooper took one look at us and said he was taking us to the hospital.

We had plenty of company there. The accident room was filled: Dillon, Martin, Mondello, Kilroy, Riordan, at least a dozen in all. None of them looked too good, but Jack Dillon was lying there gray and drawn and just barely conscious; it scared me to see him like that.

The hospital staff and the civil defense were mobilized. They processed us, taking blood tests and sending us down to the basement, where rows of cots had been set up. At least fifty firemen were down there, being monitored constantly by nurses who had stayed on duty voluntarily. They gave us breathalizer tests to determine how much toxicant we had taken in. One of the nurses told me that doctors and specialists from Harvard and MIT had convened an emergency study group to determine what possible effects the poisons in the burning plastic might have on us.

I was too exhausted to care. I lay on the cot with the room spinning whenever I dared to open my eyes. I pressed my hand over my eyes to shut out the light and stop the pounding in my temples, but the smell of smoke on my hands made me sick. I tried to sleep but was awakened by the sound of retching. The man in the cot beside me was vomiting and choking. He looked blue in the face. The nurses quickly surrounded his bed and moved him away to where they could administer oxygen.

A doctor examined me and said he wanted an EKG. One of the nurses walked me up to the accident room. In the corridor we passed the wives of two of the men; they looked upset and frightened. I took a minute to call Bev. She had heard the news on the radio and people had been calling her all day long. I told her that they were only keeping us for observation and tests, but there was no telling how long we'd be there.

"There's something wrong, isn't there?" she said.

I told her if something was really wrong somebody else would be making the call. That seemed to reassure her.

Around midnight they began releasing some of the men. They were satisfied that poisoning by the plastics was not a factor in our illness, that it was primarily a case of overexertion and severe smoke inhalation. One by one the cots in the basement emptied. By two A.M. only Davey and I and a guy from another company were left. They didn't like the results of our EKGs and were going to admit us to the wards upstairs. I had a terrific splitting headache, and although I didn't want to admit it, the pains in my chest were growing worse.

They took me up to a room. I peeled off my filthy clothes and, without bothering to wash, crawled between the sheets.

They woke me about an hour later and took me down for another EKG and a blood test. I was half asleep when they brought me back. They woke me again later in the night, but I don't know what for. I remember thinking that I was at home and Larry had come into my room wanting to play. I told him no, that he was a big boy and that I wouldn't always be here to play. But he just shook his head and climbed up on me, and we played.

=*NINE*=

WE HAD JUST COME ON DUTY FOR THE SATUR-
day-night shift when we had a call for a guy who'd been shot
on Mass. Ave. near Porter Square. We pulled up in front of a
small novelty store. Inside, the place bristled with cops; the
faintly sweet smell of gunpowder still lingered in the air.

An old man lay stretched out behind the counter, his
face very pale, the blood soaking through his pants. The
bullet had gone through his groin. Swanson used his pock-
etknife to cut away the cloth, and the old man said that it
only hurt a little bit but he couldn't feel his legs. Cooper told
him that was normal and not to worry, while Finnegan
placed a sterile gauze over the wound. Cooper told the man
to try to relax, but he wanted to talk. He had just been closing
up his store when this fellow had come in with a gun. The old
man had said, "You want the money? Take the money. It's
yours." He said he hadn't given the fellow any trouble. He
said he'd handed him the money and the fellow had just fired.

On the truck going down we gave him oxygen to keep
him from going into shock. He seemed to be aging right in

front of us: his eyes grew duller and sank deeper into the frozen mask of his face. I squeezed his hand and told him we were almost there. His hand felt cold. The old man said he was all right except for not feeling his legs.

"I didn't give that guy any trouble," he said. "What kind of a town is this?" Then he licked his lips and twisted his mouth into a sort of half-smile. "Big me," he whispered. "I must've really scared him, huh?"

He was a very old man and I don't think he knew how badly he'd been hit. At the hospital they took him straight up to the OR.

We didn't have much time to think about him. At the hospital we received a call for an elderly woman up in North Cambridge. She turned out to have suffered a stroke and we ran her down to Mt. Auburn. Joe supported her head while Dave held a sterile bowl into which she vomited all the way down. She was a thin, toothless old woman in a plain gray nightgown. No one in her family had offered to accompany her to the hospital, but she seemed too old to notice or care. She gurgled to herself like a baby in a high chair, smacking her lips and wiping her mouth with the back of her hand. When she finished vomiting Dave spread a towel over the bowl. The old woman grabbed the towel and began wiping out the bowl. Dave told her it was all right, we'd take care of that for her. I threw open the rear doors; it was a brisk October evening and as we turned up the hospital driveway a few dead leaves swirled into the Rescue.

Billy and I carried her in and when we came back out Dave had already cleaned up the old woman's mess. "Yessir," he said as we climbed aboard, "I sure am gonna miss all this. Remind me to send you a postcard." After five years on the Rescue Dave was being transferred to an Engine Company. He'd be leaving in a week or two.

Cooper was leaving, too, to become the Captain of Ladder Three. I don't think Billy really wanted to leave, but he had to accept whatever vacancy opened up. The week before, we had all taken him out to dinner at Pier Four to celebrate. We'd ordered cocktails and Cooper had begun telling Billy Stone and me the secret of studying for the Lieutenant's exam, which we were both planning to take in a year. The waitress arrived with the cocktails, and as Dave popped up to deliver the first toast he spilled half his drink in Joe's lap. They both broke out laughing while the waitress looked scandalized.

"Don't pay any attention to them," Cooper told her. "They don't get out of the house much." We all cracked up like a bunch of idiots.

Cooper never did finish telling us the secret of how to pass the exams, and by the time we got around to the toasts none of us was at his best. But Cooper must have known how we felt about him. It's easy enough for some distant authority to give orders. It's a much harder thing for a man to command respect on a day-to-day basis. Billy Cooper looked out for our welfare, and we always knew that he'd never ask us to take a chance he wasn't willing to take himself. He was a good officer and a good friend and we would all miss him.

Back at the house Eddie Kilroy was putting together a roast beef dinner that he estimated was going to cost us a buck apiece. "How about you blokes on the odd-job machine?" Eddie said. "Shall I count you in?" All of us blokes wanted in.

The English accent was the game of the week with Kilroy. A couple of days earlier an English fireman, on a visit to the States, had dropped by the house to—as he had put it—see our show. He was a member of a small country department that made as many runs in a month as we sometimes make in a morning, but we had all escorted him on a tour of the apparatus floor and right away Joe had asked him if they had a Rescue over there in England.

"A *what?*" said the Limey. With his white walrus whiskers he reminded us all of Commander Schweppes.

"A Rescue," said Joe, gesturing toward the truck.

"Ah," said the visitor, "you mean the odd-job machine." Kilroy wouldn't let us forget it.

The Englishman turned out to be a pretty decent guy. He was interested in rescue work and he spent a good part of the evening with us. We even took him along for a run. Some guy had tried to snatch a woman's purse as she was getting off a bus. She wouldn't let him have it and he had punched her in the eye with a rock. It was an awful mess. The Englishman was at first outraged; then, as we rode back to the house, he became increasingly upset. Joe asked him if they didn't have stuff like that over in England.

"Well, of course," he said, "we're starting to have what you call mugging in London, mainly in the Underground. But it hasn't spread to the country yet." Back at the house he calmed down and showed us snapshots of his home and family. He didn't earn as much money as we did, but it looked as though he lived well. He had a nice little house in a clean, pretty town. It looked like a good place to bring up kids.

He asked me if the Cambridge Rescue was typical of other fire rescue companies in America, and I told him no, that it provided a unique combination of fire, rescue and emergency medical service. We talked a little about the state of emergency service in England. The English, it seems, are way ahead of us: in London the emergency ambulances carry doctors aboard. Although a few big cities here have doctors aboard their rescue vehicles, the general trend in this country has been to upgrade the capacity of emergency personnel by training and equipping them to perform such operations as IVs, EKGs and defibrillations at the scene. But in general these programs for the different groups that perform emergency services now—hospitals, private ambulances, police or other city agencies—are only in the planning stages and progress has been very slow. We seem years away from any coordinated national plan for improved emergency service.

It seems to me that if this nation is truly committed to achieving better emergency service as quickly and economically as possible, the nation's fire departments are a perfect nucleus around which to mobilize. Throughout the country we already have a network of departments manned by people whose basic motivation is lifesaving work, departments that are already set up to provide instant response and which have a paramilitary structure capable of supervising such a program and maintaining the highest standards. The experience of the Cambridge Fire Department Rescue has proven, I think, that firemen can perform this dual role of fire fighting and emergency medical service. Not all rescue companies must also be able to fight fires; but as a practical solution to the problem of how to organize better emergency service throughout the country, rescue units could certainly be operated through the fire departments.

Around eight-thirty we had a call to go down to Cambridgeport for an asthmatic. There was no one at the door of the six-family house, but inside the hallway I caught a glimpse of a little kid in pajamas. He darted up the stairs and out of sight. One of the doors opened and we asked the woman who appeared if someone had called the Rescue. She said she didn't know.

A woman's voice yelled, "Up here." We ran up to a second-floor apartment. Inside the front room a woman in her late thirties lay stretched out on the sofa. Behind her, climbing onto a seat at the table, was the little kid in pajamas.

"Take me to the hospital," the woman said, pushing herself up off the couch and clutching her robe around her. "I wanna go to the hospital."

"What seems to be the problem?" Cooper said. "You having difficulty breathing?"

"Yeah. I'm having difficulty."

Cooper and Joe went up to examine her. Her eyes were slightly bloodshot.

"What're you doing?" she said.

"Don't worry," said Billy. "We just want to check you over."

"You don't have to look at me none," she said. "I already told you what's wrong. I wanna go to the hospital."

I moved closer. She wasn't having an asthma attack, that was for sure. There didn't seem to be anything wrong with her except for the booze. Up close you got a good whiff.

Cooper made a show of checking her eyes, her pulse, her respiration. At the table behind us the little kid was playing with bottle caps. He had made a goal out of the salt and pepper shakers and he was shooting the caps into it.

"Well," Cooper said, completing his examination, "I think you'll be all right now. Is there anything else you need us for?"

"Is there anything else?" the woman repeated in disbelief. "You betcha there is. I want you to take me the hell outta here."

"Just where would you like us to take you?" Cooper asked her.

"Whaddaya, hard of hearing? I already told you. I wanna go to the hospital." She was shouting now and every time she opened her mouth the smell of liquor washed over us.

Cooper said he was sorry, but we couldn't take her to the hospital without a reason.

"All right, then," she said, pulling her robe tight around her, "he hit me."

We all looked over at the little kid, who was still shooting bottle caps.

"Not him, the other one," said the woman, gesturing toward the rear of the apartment. "He's hiding in there. He's afraid to come out."

A man's voice hollered, "I did nothing of the kind. I didn't do a thing."

From where I was standing I could see the guy in the back room reading the papers. He just looked up at us, moistened his finger and turned the page.

"He did," said the woman, spreading open the collar of her robe. "Look what he done to me. He tried to choke me." There wasn't much to see. Her neck looked a little reddish. "He tried to choke me and then he tried to throw me over the back porch."

"Don't listen to her," said the guy from the other room. "She doesn't know what she's talking about."

"He did," shouted the woman, her eyes brimming now with tears. "He hit me."

"She's drunk," said the man. "She doesn't know what she's saying. She's always drunk."

"Shut up, you heel," she snarled through her teeth. She grabbed hold of Cooper's wrist. "So what if I take a drink? Is that a crime? But he tried to kill me. He said he was gonna throw me off the roof. Please, you gotta take me outta here or he'll kill me."

Cooper told her we'd call the police if she wanted, but that we couldn't take her to the hospital unless there was something wrong with her.

"I got asthma," she said.

"You seem all right now," he said.

"I wanna go," she said. "He tried to hit me."

Cooper looked up at us, shaking his head in frustration. There was nothing the matter with the woman; it was just another family quarrel and a question of deciding whether to call the police, do nothing or take her to the hospital just to get her away and give them all time to think. The woman was sobbing now.

"Hey, you guys," called the man from the other room. "Don't waste any more of your time with her. I never laid a hand on her."

Suddenly a volley of bottle caps pelted off the wall. "Yeah, you did too," hollered the little kid, climbing down off the chair and moving toward the man yelling, "You did too hit her! You did too!"

"All right, son," said Cooper, putting a hand on the boy's shoulder to quiet him. Billy looked up as if to ask what we thought. I guess we all felt it was worth taking her out of there for the kid's sake.

"Sure, take her away," said the man. "You'll be doing me a big favor."

The woman stared blankly ahead as we strapped her into the chair. When we took her out she didn't say anything—not even good-bye to the kid. We left him alone in the room picking up the scattered bottle caps.

We got back to the house just as Kilroy's dinner was getting underway. Hungry men crowded around the long table waiting for their plates; those who had been served were shoveling it in as if they were competing in an Olympic event. At the head of the throng stood Kilroy, carving the meat; Jack Dillon and Jerry Martin were on either side of him, dishing out the mashed potatoes and the vegetables.

"Well, blimey, look who's here," said Kilroy as we took seats near him. "If it isn't those odd-job-machine blokes. Well, OK, you odd blokes, sit down and have your fill. Take your time. Then when you're through would you mind terribly much popping up to the Square. There's an Italian who hurt himself raking leaves—fell out of the tree. You'll recognize him well enough. He'll be waving a bloomin' salami over his head."

"Eddie," I said, handing him my plate, "did I ever tell you about this drunken Irish fellow I ran into one night?"

"Ooooh, Larry," said Kilroy, carving with great flourishes, "you're not calling me ancestors drunkards now, are you?"

"Honest to God, Eddie, this is the first one I ever saw under the influence. That's why I remember this particular story. It was very sad. You see, he had lost an eye in the war, and being as he was in the Irish Army and they couldn't afford glass, they had to give him a wooden one. You know how it is."

"I do indeed," said Kilroy. "He's lucky they didn't give him a potato eye and a pat on the back."

"Yeah, well," I said, "naturally he was very self-conscious about it. For a long time he wouldn't go out at all. Then one day a friend came by, told him he was crazy to ruin his life over a little defect and talked him into going to this dance. The Irishman had himself a few drinks, just to build up a little courage, but when he got to the dance hall and saw all the girls he began to lose his nerve. So he said to himself, I'll see if I can find a girl with an affliction, then it won't be so bad. So he made his way through the hall, carefully looking over all the women, until way over in the corner, sitting against the wall all by herself, he saw this girl with a harelip. So he straightened his tie, walked right up to her, bowed, cleared his throat and said, 'Excuse me, miss, would you care to dance?'

"The girl sprang to her feet, saying, 'Would I? Would I?'

"And this guy, he took a step back and shouted at her, 'Harelip! Harelip!' "

Kilroy laughed easily. "All right," he said, laying a couple of juicy slabs of roast beef on my plate, "you've earned your dinner. Too bad we don't have any pasta for you, though," he said, passing my plate to Jack Dillon. "Jack, easy on the potatoes. Larry's not used to them."

"Listen," Jack said, "with what potatoes are costing these days I'm taking it easy on everybody."

"You ain't kidding," Jerry Martin said. "A bag of spuds, sixty-nine cents. Can you believe it? Eggs a buck twenty a goddamn dozen. Chicken ninety-eight a pound. At home we used to have liver and onions once a week. Now you go into the Star Market and you can't even get it. The other day I says to the butcher, 'Well, what's the special?' He says, 'There is no special.' 'Well,' I says, 'what's a good buy, then?' 'There isn't a good buy in the store,' he says. I tell you. It's that goddamn Nixon. He's ruining the country. He doesn't care if we can't eat."

"Or keep warm," Jack Dillon added.

"Fuel shortage," said Finnegan, wolfing down a mouthful of beef. "There's no fuel shortage. What there is is a bunch of big oil companies saying whatever they hafta in order to raise prices. I don't understand what's going on down in Washington. Where're all the honest politicians?"

"Honest politicians?" Dave said. "Any politician important enough to swing any weight knows where his bread is buttered. They're all in with big companies in one way or another. They make a lot of noise, but what they're fighting for is to see whose companies get a bigger piece of the action. And meanwhile the rest of the country is going to pot."

Jack Dillon said, "You don't have to look at Washington if you wanna know what's wrong with this country. You can see it happening right here. People aren't together anymore, that's what's wrong. They don't talk to each other. Jeez, in the old days, in the spring and summer, everybody used to be out on the porches talking. You don't see that nowadays. People just don't have the time, and when they do stop long enough to talk, they don't have the conversation in them, because they're all filled up with the TV. Nowadays you go out on the street and everyone is strangers. It's like living in an apartment, or in New York."

Kilroy was on his feet, banging his tonic bottle with a knife. "Hey, what're you guys trying to do? Turn this dinner into a wake?"

"Hey, no, it's terrific, Eddie," Cooper said, holding up his coffee cup. "Whatever deficiencies you may have as a person, you make up for them with your cooking."

"Thank you veddy much, Guv," Eddie said. "That's veddy kind." He raised his tonic bottle. "Hey, fellows, how about a toast here for the Commander of the odd-job machine. And a regular bloke he is. Calls us by our Christian names off the job, 'e does. We're gonna miss you around here, Guv. Now you and the odd balls take your time, Commander. Then when you've had plenty of time to digest your vittles, would you mind terribly much popping up to the Square? There's a casualty situation: ten Italian workmen got dizzy changing a streetlamp."

"Don't tell me," I said. "One of them held the bulb while the other nine turned him around."

"Oh, you must know those guys, Lar," Kilroy said.

"We can't make it up to the Square just now," said Dave, and he started to tell the one about the Christmas lights imported from Ireland, when the loudspeaker came on: *"Attention. Rescue going out."*

We jumped up from the table and hit the poles. On the apparatus floor Bob Mondello, who was on patrol, called out, "Report of a bad accident on Fresh Pond Parkway in front of the shopping center."

"Bad accident," said Dave as the Rescue roared out. "They always say bad accident. Is there such a thing as a good accident?"

"I dunno about that," said Joe, rubbing his stomach, "but there's such a thing as eating too fast."

Billy Stone was at the wheel for the long run up Concord Ave. The screaming of the siren and the red lights flashing off the corridor of houses gave the peaceful autumn night the ugly, garish quality of a discotheque.

Ahead of us a long curving line of red brake lights was strung out like some strange religious procession. The two right-hand lanes were completely blocked with traffic. As Billy Stone pulled out into the oncoming lanes, we saw the spinning blue lights of the cop cars and, facing each other head on across the middle of the road, the two wrecks. Hoods were up, doors thrown open, windshields shattered, glass strewn everywhere, and dark stains spreading across the asphalt. A huge crowd—literally hundreds—lined the sides of the road.

As the Rescue pulled to a stop we jumped off and ran to the cars: they were both empty. "Everybody's all right," hollered a cop. "The kids're drunk. We got them in the squad car."

"You sure everybody else is OK?" said Dave.

"Well, some guy over there was driving one of the cars, but he's not hurt," the cop said.

We went looking for the guy. You never take anyone's word until you see the victim himself. We spotted the guy standing at the edge of the crowd. He was a heavyset man in his early forties. His shirt had come untucked and he was pressing a hand to his chest. He looked apprehensive as we approached. Dave asked him how he was feeling.

"Oh, you know," he said. "Nothing serious. I just got a little shook up. My chest hurts a little, that's all."

As soon as he said that, Billy Stone went for the chair. People at the scene of an accident want desperately to believe they're all right, and they may look and act all right, but if you've never seen them before you have no way of knowing how they normally behave. For all you know they may be in shock. You have to take every precaution.

While we were waiting for the chair the man kept insisting he was OK. He began getting upset.

"We know you're all right," Dave said, patting him on the shoulder, "but we're required to check you out for the record. It'll only take a minute."

"OK," the man said, letting out a deep breath. He winced and clutched at his side. Dave asked him where he hurt. He shook his head and took a step away from us.

Billy brought the chair and unfolded it. The guy didn't want to sit down, but Dave told him it was just part of the procedure and eased him into it. We strapped him in and wheeled him to the Rescue, where Joe helped us lift him aboard. He and Cooper had checked over the two kids. "They're not exactly sober as judges," Joe said.

On the way down the man held his chest, his face taut, his lips compressed. He was a big man, and squeezed tightly into the chair he looked like an overgrown kid.

"How old are you?" Davey asked him.

"Thirty-nine," he said. "No, forty." He put his fingers to his mouth and rolled his tongue around. There was blood on the inside of his mouth and on his fingers when he took them away.

"It's just a little cut," Dave told him.

"I know," he said, his eyes moist. "Is everybody all right? Are the others all right?"

"They're fine," Dave said. "They were just shook up."

"I hope those kids are all right," he said. "Jeez, they were just young kids. I gotta call my wife right away. I got three kids myself," he said, reaching back for his wallet. He winced again and grabbed his side. It looked to me as though he'd had a few drinks himself.

"What's your name?" Dave asked him quickly.

"Al," said the guy, his mouth working a little at the corners.

"Al what?"

"Al Martinelli," he said. "I'm a tough sonofabitch." He gripped Dave's hand and started squeezing it to show how strong he was. Dave squeezed back. "I really am," the guy said, releasing his hold.

"Well, so am I," Dave said.

"I know," the guy said. "I was lucky to get you."

He raised his hand to pound Dave on the shoulder, but he had forgotten about his chest and the pain must have really shot through him. He cried out, his expression changing instantly to one of helplessness, his eyes welling suddenly with tears.

"I want you to stay with me," he begged. "Please stay with me."

"Sure, no problem," Dave said.

The guy had been lucky. The two drunk kids in the cop car had been lucky, too, although they probably didn't feel very lucky at this moment. Traffic accidents cause over fifty-six thousand deaths a year. It's estimated that the use of alcohol leads to over half these accidents. Every ten minutes there's a motor-vehicle death. Every sixteen seconds there's an injury. Just like a clock, day and night, day in and day out. There's no end to it.

When we returned to quarters Dave and I stood for a while out on the apron smoking cigarettes. The crisp air smelled of fallen leaves, and the moonlight silvered the roof of the church in the Yard. I like autumn best of all the seasons; it's fresh and invigorating. It was hard to believe another year had passed.

The light of the red lamp on the front of the house tinted Dave's face as he spoke; it looked as though he had a slight sunburn. He talked about some of the weekend trips we had made with other firemen and their wives, to the mountains for skiing and to the Cape in the summer, and how we'd have to make an effort to keep them up now that he and Cooper would be stationed in different houses. We'd have to keep the old company together, Dave said. After all, he said, we were rescue men.

"Some rescue men," I said, and we both laughed, remembering a time last summer on the Cape.

We were having dinner with five other couples at a long table. Everybody was laughing and drinking and eating and all of a sudden one of the girls turned red and started gasping. We all watched, frozen to our seats, as she began to choke. I saw Dave sitting there and suddenly it dawned on me that I knew what to do. It must have hit him then, too, because we both jumped up at the same time.

"What's the matter?" I said, grabbing the girl, and before she could answer I tipped her forward and whacked her once on the back, dislodging a piece of meat. A minute later the girl was fine, and for the rest of the meal the others were riding us: two rescue men at the table and neither of them had exactly sprung to the rescue. But long after the meal was over I couldn't forget that instant of astonishment when I realized that I wasn't just another person: I was a rescue man.

Dave pulled up his collar, then said it was too cold out and went indoors. I stayed outside to finish my cigarette. It was still a fine, crisp evening. A Jaguar XKE came purring across the apron. The driver, a long-haired kid, wanted to know how to get to Mass. Ave. I asked him where on Mass. Ave. and he said, "Porter Square; we're going to a party there." As I bent down to give him directions I saw the perfectly beautiful dark-haired girl on the seat beside him. She smiled to herself, he thanked me and the car pulled out with a deep, throaty roar.

For a long moment I wished very much that I were twenty-one, with a pretty girl beside me, heading off to a party somewhere in the night, feeling that all the world was a path to something better and that I would live forever. I tried to remember what it had been like to ride past a firehouse without giving it a second thought.

After a while I went in and watched the late news with Finnegan and Billy Stone and Bob Mondello. Around eleven-thirty a box came in: Box 321, Clinton and Harvard streets, just around the corner. Everything was going.

The Rescue shot out of the house. We pulled on our gear fast. In under a minute we were turning onto Clinton Street. There was a thin haze of smoke on the street; we could smell it. Engine One was right behind us as we stopped in front of a three-story frame house. The smoke was coming out of a second-floor window.

We ran up to the second floor. Smoke was seeping out from under the door and we put on our masks. Dave ripped off a glove and said the door was cool. Cooper sent Joe and Billy Stone upstairs to evacuate.

We tried the door but it was locked. Dave jammed in the halligan bar and popped the lock. Dropping to our knees, we pushed open the door. I heard a faint cry and jumped back, startled, as a cat streaked down the stairs. Then I was following Cooper and Dave into the smoke. We moved in different directions, looking for the fire, but there was no red anywhere.

I crawled straight ahead into what I guessed was the living or dining room. I heard a crash, then the tinkling of broken glass. I slammed into something, a chair, reached across it—and felt the edge of a bed. My arm swept the mattress, and suddenly, as in a dream, it met resistance.

"I've got someone," I cried out. Leaning close through the smoke I made out the features of a man: he was still breathing. I shook him and he mumbled, as Davey grabbed hold of me from behind. The two of us picked him up and carried him out into the hallway.

We sat him down on the stairs and pulled off our face-pieces. He was about fifty-five and he started coming out of it as we checked him over: his eyes, pulse and respiration were all good.

"You handle him," Dave said, and he went back to the fire. The guy was coughing now and trying to shake me off.

"What the hell's going on?" he mumbled out of a half-drunken stupor. "What the fuck're you doing?"

Joe and Billy Stone were coming down the stairs. "Get this guy some oxygen," I said. They took him down with them. I could hear the Engine Company coming up the stairs as I went back into the blackness of the apartment.

There was still no red showing. I made my way back into the room where I'd found the man sleeping.

"Who's there?" Cooper called.

"Larry."

"I've found it," he shouted. "It's in the closet. Get a line in here."

Out in the hallway Dillon, Kilroy and Kelley were making the bend up the stairway with an inch-and-a-half. I led them into the bedroom and they hit the closet. Cooper, Dave and I went through the place opening windows and looking for people, but that was it.

The Deputy came in to examine the closet. He sent a couple of guys downstairs to check the extent of the fire, then he set two of the laddermen, Martin and Mondello, to clearing out the charred remains of the closet. The clothing had been reduced to a wet, steaming mound of black ashes on the floor. The fire must have been smoldering in there for hours; another minute or two and the guy could easily have died from smoke inhalation.

At the far end of the smoky hallway one of the men was staring at me from under his helmet. I stared back, certain that I knew him, but unable to place his name. Then Cooper was calling for me in the other room. The instant I moved, the guy who was watching me moved. For Chrissake, I thought, I'd been looking at myself in a mirror.

"Say, Dep," called Cooper, "we're taking that guy down to have him checked out."

"All right," said the Deputy, waving a white-gloved hand at us. "OK on the Rescue. And thanks."

We made our way out into the haze of blue smoke that hung in the street. The crowd of onlookers was beginning to disperse. Joe and Billy Stone were in the back of the Rescue with the man. The guy was feeling his oats. He wanted to know where we were taking him. He wanted to know what all those people were doing in his place. He wanted us to know we'd have to pay for anything we broke or stole. He wasn't gonna let us walk over him, no sir, whoever the hell we thought we were.

"You're gonna pay for this," he said, as we shut the doors and pulled out for the hospital. Then he started coughing badly.

"All right, Pops," said Joe, giving him a shot of oxygen, "why don't you take a breather now."

It was after one o'clock in the morning when we returned to the house. We made coffee, then washed up and sacked out. From the window beside my bed I could see the steeple in the Yard. A gust of wind rattled the windowpanes and sent a flurry of leaves swirling to the ground. The trees were almost bare now. Another year was ending.

Finnegan and Billy Stone were sleeping soundly; Dave tossed uncomfortably. I lay in bed and wondered if that young couple had made it all right to their party. I tried to remember back to when night in this city was a time for parties and excitement, and not the things I had come to know it for.

In the middle of the night, I still wonder sometimes if I am doing the right thing. To reach out, to help people, to find someone on the verge of death and be able to bring them back—for me that's a miracle. But how often can you come face to face with the endless fact of pain and death without losing courage in life?

In my first year on the Rescue we picked up a man whose face I've never forgotten. We'd had a call to go to a building under construction, for an injured man. As we descended into a darkened subcellar I noticed a strange electrical smell in the air. The workmen at the base of the stairs said that the man was in the middle of the floor, but the cellar was empty except for a few big electrical boxes on the wall and a black man leaning casually against a sawhorse. I scanned the floor for a body, then, looking up at the black man, realized suddenly that he was the injured man; that his belt and his work boots were smoldering, and his clothes burned off. Then I saw that all his hair and his eyebrows were gone as well. His flesh was charred black from head to toe. He was completely black except for the whites of his eyes.

The shock must have shown on our faces because he said, "Yes. It's me." We laid sterile sheets over him, planning every movement carefully, knowing that the instant we grabbed hold of him the burned flesh would break off. We wrapped the sheets around him and put the orthopedic stretcher up against his back where he stood, and then gently lowered him onto it.

On the way down he just lay there looking up with those big eyes. His eyes seemed to grow enormous, as if all the remaining life in him was concentrated in their whiteness. He lay there without making a sound and I smelled the faint odor as the sheets soaked through. With his skin gone there was nothing to contain his body fluids. I knelt over him, telling him he was going to be all right, and he said to me, as clear as a bell, "I'm not going to be all right. I've been in this business too long. I know how much juice I took," he said. "I'm going to die."

I had just come on the Rescue then and I could only think to myself: He doesn't know how badly he's hurt. He must be in shock; he doesn't really know what he's saying.

Only much later could I accept the fact that when he spoke to me he was perfectly rational. He knew that he would die, and a few hours later he was dead.

Even now I can feel the steadfastness of his gaze. If he could find the courage to face certain death, surely I can muster enough to face this life.

But I try not to think of ghosts. This town is made up of living people, some of whom may need our help. There are even a few people in this town alive today because of us, and there are many others whom we've helped to avoid a more serious injury and unnecessary pain. I try to think of them.

The other day Finnegan showed me a two-page close-up photograph of a tropical butterfly. The colors were brilliant and the pattern incredibly complex. It was one of those pictures that suddenly force you to see something in a way you've never seen it before. Joe asked me if the picture reminded me of anything and I said no; I'd never seen anything quite like it. "No, c'mon," he said, holding open the book and stepping back. "What does it look like?" He took another step back and I saw what he meant. In the outline of the butterfly he had seen a map of Cambridge.

This town could be a fine place to live. I lay there thinking about all the ways it could be better, trying not to listen for the next alarm, trying not to think of ghosts. Thinking of my wife, my children. Thinking of sleep.

At daybreak we were awakened by a call for an injured man. We pulled up in front of a barroom off Mass. Ave. A cop, rubbing his hands together in the cold gray half-light, said the guy was out back.

The alleyway stank of garbage. We found the guy lying in the trash that had overflowed the barrels. He was ageless—fifty going on a hundred; a real grubby drunk, his face covered with scabs and a three or four days' growth of stubble. He lay shivering in his ragged overcoat, his arms wrapped around himself.

We bent over him. There was a gash on the back of his head and his teeth were chattering from the cold. We wrapped a blanket around him. He looked straight up at us out of the two red slits that were his eyes and said he couldn't take any more of it. He said he had to give up the booze. He said he'd had enough.

On the way down in the Rescue I searched his face for some sign that he'd meant it, wondering how anybody ever knew when they'd had enough. I don't know if he ever quit; but I'd like to believe that he tried.

Appendix

What You Can Do in Emergencies: A Capsule Guide

Note: This guide is not intended as a substitute for first-aid training. Its purpose is only to provide you with recommended courses of action for the most common emergency situations.

FIRE

Call the Fire Department. This may seem self-evident, but many people neglect to do it, assuming that they can fight the fire themselves. Fire is trickier than most people realize. Hours after you think you have extinguished a fire in the wall behind your stove, or in your mattress, it can burst into flames again. Firemen are trained to understand fire and to respect it. No fire is too small; call the Fire Department.

When fire strikes, *get everyone out immediately.* Forget your valuables. Close all the doors behind you as you leave; this will buy extra time.

If you are awakened at night by fire or smoke in a closed room, drop to the floor. Any oxygen in the room will be closest to the floor. Crawl to the door and feel it. If it is hot, do not open it: the fire outside will overtake you. If the door is not hot, open it cautiously, being prepared to close it quickly if there is fire or smoke outside. If you can't escape safely through the door, go to the window and open the top and bottom slightly. You will be able to get air through the lower opening and wait for the arrival of fire personnel.

If your door is open, and there is smoke and fire in the hall, close the door and proceed to the window as described above. Remember that smoke alone causes the majority of fatalities in fires. Be sure to explain this to children as soon as they are old enough to understand. Most of the time we find kids hiding under beds or in closets because they don't know where to go.

When you go to any place of public assembly such as a theater, restaurant or nightclub, take a second to *look for the nearest emergency exit* and fix it in your mind. When people panic they flock to the one exit they know: the door through which they entered. Fire fighters find unburned bodies literally stacked against the main door, when only a few feet away there is an emergency exit that would have led the victims to safety.

If you are caught by fire in a hotel, remember that Fire Department ladders do not reach above the sixth or seventh floor. If you cannot get to a fire escape, you should open your window from both top and bottom, soak your mattress by using an ice bucket or other container to convey water from your sink, and place the wet mattress up against the door to your room. Such measures will gain you the time it takes fire personnel to reach you.

AUTOMOBILE ACCIDENTS

Do not remove the victim from the automobile unless the car is on fire or there is some obvious danger, such as a tree or telephone pole falling on the car. Moving the victim can only add to any injuries he has already sustained. Many people have been permanently disabled or have even died as a result of being moved unnecessarily at the scene of an accident. There is a popular misconception, spread by Hollywood, that all wrecked cars burst into flames sooner or later. This is simply not true. If there is a gasoline spill, it is wise to stay away from the car and to tell onlookers to extinguish all cigarettes. If you can, turn off the ignition switch, even if the motor is not running; if you can't reach the switch, try to disconnect the battery cables. But these are only precautions. Do not remove the victim from the automobile. All first aid should be administered right in the car.

Take charge. Make sure the police or Fire Department are notified. The arrival of emergency personnel is often delayed simply because everyone at the scene of the accident assumes that someone else has made the necessary call.

If you are unable to gain access to him, at least *try to keep the victim calm.* Tell him that help is on the way. Never tell him the extent of his injuries; any mention of his injuries, however slight, may send him into shock. You can even shock a victim by suggesting that he may have to wear a cast for a few weeks. Tell him he'll be all right.

If you are able to gain access to the victim, check to see that he can breathe. Do not move his head, but check to make sure that his airway isn't blocked by dentures or other foreign matter, and that his tongue hasn't fallen back into his throat. Use common sense in clearing an airway. Whatever you do, don't move the victim's neck; to do so may cause irreparable damage.

If the victim is bleeding profusely, apply direct pressure to the wound. If possible, use a sterile cloth; if not, place your hand directly over the wound.

If the victim is uncomfortably jammed up against the dashboard, use the seat release to move the seat back. Under no conditions should you try to move the victim bodily.

If the victim is apparently not seriously hurt, use your common sense in restraining him from unnecessary movement. People want to believe they are uninjured. If there is even the slightest indication of a possible injury, try to persuade the victim to remain still just as a precaution.

Keep onlookers back. They can only upset the victim and impede the work of the emergency personnel when they arrive. Clear a path for the emergency personnel and keep it clear. When they arrive, give them a brief evaluation of the situation. Tell them who is where with what injuries.

STREET EMERGENCIES

The procedures are similar to those for automobile accidents. Make sure that a medical emergency unit has been called. Keep the crowd back. Keep the victim calm; assure him that help is on the way and that he will be all right. Do not discuss the extent of his injuries. Keep a path clear for the emergency crew, and give them a brief report of the situation.

EMERGENCIES IN YOUR HOME OR PLACE OF WORK

Keep a list of emergency numbers near your phone. In the pressure of an emergency you, or your child, may be too upset to perform simple tasks like looking up telephone numbers, and valuable minutes may be lost.

When you call for help, *try to stay calm.* Speak clearly and give only the most relevant facts: the precise address and the nature of the emergency.

If possible, *send someone outside* to meet emergency personnel and lead them directly to the patient or victim.

If you are in a building with an elevator, have someone *hold the elevator* on the ground floor.

When the emergency personnel arrive, give them as many relevant facts as you can. In the case of an accident, tell them what happened, how long ago and what has been done. In the case of illness, tell them how long the patient has been ill and what medication he has been taking. If possible give them any medication or chemical involved to take with them to the hospital.

Let the emergency personnel decide which hospital is most appropriate. Inform them if the patient's records are kept, or his doctor practices, at a particular medical center, but do not insist that he be taken to that place. The emergency personnel have to consider traffic patterns: getting the patient to a hospital *quickly* can save his life.